THE COOPERATIVE EXTENSION SERVICE

Other Titles in This Series

†*Rural Society in the U.S.: Issues for the 1980s,* edited by Don A. Dillman and Daryl J. Hobbs

Rural Public Services: International Comparisons, edited by Richard E. Lonsdale and Györgi Enyedi

The Impact of Population Change on Business Activity in Rural America, Kenneth M. Johnson

†*Science, Agriculture, and the Politics of Research,* Lawrence Busch and William B. Lacy

Technology and Social Change in Rural Areas, edited by Gene F. Summers

The Social Consequences and Challenges of New Agricultural Technologies, edited by Gigi M. Berardi and Charles C. Geisler

†Available in hardcover and paperback.

About the Book and Authors

The Cooperative Extension Service:
A National Assessment
Paul D. Warner and James A. Christenson

The Cooperative Extension Service, a publicly supported educational agency, is continually struggling to define its proper function and purpose in our changing society. Should its mission be broadly based or narrowly focused? Should staff members be generalists or specialists? Should its clients be primarily rural or urban, farm or nonfarm? What role should Extension play in the information networks of the twenty-first century?

Professors Warner and Christenson take a broad look at these and other questions concerning where the Extension Service has been, how well it is doing, and where it ought to go. Theirs is, first, the only comprehensive national survey that looks at the total Extension organization rather than at just one program area. Second, it expresses the viewpoint of Extension clients and the public, rather than that of the organization's staff; and third, it combines outside survey information with data recorded in the Extension Management Information System (EMIS) and other routine agency reports.

The authors evaluate, among other things, the extent of public awareness of the agency and its four major program areas (agriculture, home economics, 4-H, and community development), determine the users and nonusers of the programs and the accessibility of programs to the general population, identify the level of satisfaction with existing programs, and outline priorities and policy issues for the future.

Dr. Paul D. Warner is professor of sociology at the University of Kentucky and Extension specialist with the Kentucky Cooperative Extension Service. He has conducted several evaluation studies of innovative program delivery systems, the most recent of which explored the organizational impacts of a videotext information system for delivering weather, market, and production information to farmers. **Dr. James A. Christenson** is professor and chairman of the Department of Sociology at the University of Kentucky. From 1972 to 1976 he was Extension specialist at North Carolina State University. Since 1979 he has served as director of the Survey Research Center at the University of Kentucky and is currently the editor of *Rural Sociology*.

THE RURAL STUDIES SERIES
of the
Rural Sociological Society

Editorial Board

Chairman, David L. Brown, ERS, USDA, Washington, D.C.

Jere L. Gilles, University of Missouri
Denton E. Morrison, Michigan State University
Sonya Salamon, University of Illinois
Marta Tienda, University of Wisconsin
Kenneth P. Wilkinson, Pennsylvania State University
James J. Zuiches, Cornell University

THE COOPERATIVE EXTENSION SERVICE: A NATIONAL ASSESSMENT

Paul D. Warner and
James A. Christenson

Westview Press / Boulder and London

Rural Studies Series, Sponsored by the Rural Sociological Society

All rights reserved. No part of this publication may be reproduced or transmitted in any form or by any means, electronic or mechanical, including photocopy, recording, or any information storage and retrieval system, without permission in writing from the publisher.

Copyright © 1984 by the Rural Sociological Society

Published in 1984 in the United States of America by Westview Press, Inc., 5500 Central Avenue, Boulder, Colorado 80301; Frederick A. Praeger, President and Publisher

Library of Congress Catalog Card Number: 84-50486
ISBN 0-86531-752-6
ISBN 0-86531-753-4 pb

Composition for this book was provided by the editors
Printed and bound in the United States of America

10 9 8 7 6 5 4 3 2 1

To our parents,
Ruth M. and Lorraine V. Warner
and LaVerne D. and Gilbert R. Christenson,
who shared with us
their values, love, and appreciation of life.

Contents

List of Tables and Figures xiii
Acknowledgments xvii

    Introduction 1

1   **Extension in Changing Times** 5

    Extension's Roots Go Deep 6
    A Unique Organization................................. 10
    An Unprecedented Funding Arrangement................. 14
    Measuring Extension's Effectiveness..................... 16
    Establishing Legitimacy................................ 19
    Organizational Stereotypes: Myths or Truths? 20
    Issues Facing Extension................................ 24

2   **Evaluation in Extension**............................. 26

    Organizational Effectiveness Examined.................. 27
    The Input-Output Approach 29
    Systems Effectiveness Model for Extension 32
    Sources of Data 40
    Limitations of the Study............................... 40
    A Plan for Evaluation................................. 41

3   **Public Awareness of Extension**..................... 43

    The Consumer and the Cost-Bearer 44
    Public Opinion and Organizational Image................ 44
    The Image of Extension 46
    Extent of Awareness 47
    Who Is Aware of Extension?............................ 50
    Awareness of the Four Program Areas................... 51

Profile of Knowledgeables 52
Shoring Up Extension's Image 53

4 Who Are Extension's Clientele? 56

Who Uses Extension? 58
Individual and Household Use........................... 59
Regional Use Patterns..................................... 61
Rural-Urban Use.. 61
Equity in Service Delivery................................ 63
Profile of Users... 66
Clientele and Political Activity 66
Use of the Four Program Areas......................... 66
Frequency of Use.. 69
Methods of Use 69
Dilemmas of Whom to Serve 70

5 Some Clients Are Satisfied and Some Are Not 73

Subjective and Objective Performance Measures............ 73
Client Satisfaction 74
Satisfaction: A National Perspective 76
Satisfaction: A State Perspective........................... 78
Users and Nonusers.. 79
How Frequency of Use Affects Satisfaction................. 81
Issues of Service Assessment 83

6 Support for Extension 86

Clientele Involvement 87
Autonomy of Extension 87
Local Interests... 88
Nature of the Constituency............................... 89
Experience Base 90
Historical Support Base 90
Support for Extension in the U.S........................ 91
How Satisfaction Affects Support 95
Support: A State Perspective............................. 96
Issues of Support for Extension 98

7 The County as the Focus of Service Delivery 100

The Historical Context of County Programs................ 101
A County-Based Effectiveness Model 102
Demonstrating the Model with Kentucky Data 106

Contacts and Methods 109
Cost Effectiveness in County Programs 109
The County Effectiveness Model 110
County Summary....................................... 112

8 Summary and Priorities............................. 115
Awareness, Service, and Support....................... 115
Summary in Brief...................................... 120
Future Program Priorities............................. 120
From the Present to the Future 124

9 Policy Issues Facing Extension..................... 125
The Mission: Should It Be Broadly Based or
  Narrowly Defined?................................... 125
Structure: Should Staff Be Specialists or Generalists?........ 129
Resources: What Is the Best Method of Federal Funding? ... 133
Environmental Changes: What Is Extension's Response?..... 133
The Images of Extension: Can Extension Afford
  Multiple Identities? 135
Urban and Rural Clientele: Who Should They Be?.......... 137
Equal Opportunity: How Do the Disadvantaged
  and Minorities Fare? 138
Use: Try It, You'll Like It 139
Extension Methods: Personal or Impersonal? 140
Support: Who Will Champion Extension's Cause?.......... 142
Evaluation: What Is and What Ought to Be? 144
The Book's Intent..................................... 146

References... 149
Appendix A: The National Survey 157
Appendix B: Additional Findings 175
Index 189

Tables and Figures

Tables

1.1	Distribution of Client Contacts by Program Area, 1982	10
1.2	Distribution of Staff by Type of Position, 1982	11
1.3	Distribution of Staff Time by Program Area, 1982	12
1.4	Number of Extension Personnel by Type, 1914–1982	13
3.1	Knowledge of Extension by Place of Residence	50
3.2	Knowledge of Extension by Income	51
3.3	Knowledge of Extension by Age	51
3.4	Knowledge of Extension by Race	52
4.1	Individual and Household Use of Extension	60
4.2	Geographical Distribution of Users and Nonusers	62
4.3	Equity for Users and Nonusers	64
4.4	Socioeconomic Characteristics of Users and Nonusers	65
4.5	Political Involvement and Ideology of Users and Nonusers	67
4.6	Characteristics of Users of the Four Program Areas in 1981	68
4.7	Frequency of Use by Kentucky Extension Clientele	70
4.8	Methods of Communication Utilized by Clientele	71
5.1	User Satisfaction with Government Service Agencies	75
5.2	Satisfaction with Extension in Kentucky	78
5.3	Satisfaction of Users of Kentucky Extension by Personal Characteristics	80
5.4	Satisfaction of Nonusers of Kentucky Extension by Personal Characteristics	82
5.5	Satisfaction with Kentucky Extension by Frequency of Use	83
6.1	Trend in Characteristics of House Districts, 1966–1976	91

6.2	Users Wanting Less Spent by Region of U.S.	94
6.3	User Support by Level of Satisfaction with Extension in the U.S.	95
6.4	User and Nonuser Support for Kentucky Extension by Level of Satisfaction	96
6.5	Support for Kentucky Extension among Users and Nonusers	97
6.6	Support by Frequency of Use of Kentucky Extension	97
7.1	Distribution of Staff Time to Different Educational Methods, Kentucky Counties, 1977–1979	108
8.1	Summary of Major Findings	118
8.2	Distribution of Professional Staff Time by Program Component, FY78 to FY82	121
8.3	Priority for Extension Programs by Users	123
A.1	Approximate Sampling Tolerance (95 in 100 Confidence Interval)	159
A.2	Characteristics of Respondents from 1982 National Survey of Adults (18 Years and Older) Compared with 1980 Bureau of Census Data	161
B.1	Awareness of Program Areas and Organizational Name by Personal Characteristics	176
B.2	Profile of Knowledgeables of Extension	177
B.3	Users and Nonusers of the Four Extension Program Areas in 1981	178
B.4	Satisfaction with Extension by Characteristics of the Respondent	179
B.5	Satisfaction with Program Areas by Personal Characteristics	180
B.6	Support for Extension by Personal Characteristics	181
B.7	Support for Agricultural Programs by Personal Characteristics	182
B.8	Support for Home Economics Programs by Personal Characteristics	183
B.9	Support for 4-H Programs by Personal Characteristics	184
B.10	Support for Community Development Programs by Personal Characteristics	185
B.11	Correlation Coefficients for County-Level Variables in Kentucky	186

Tables and Figures   xv

B.12   Number of County Staff Contacts (per capita) and Cost per Contact Regressed with Educational Methods and County Situation Variables 187

Figures

1.1   Farm and Rural Population, 1910–1980 8
1.2   Ratio of County Staff to Specialists 14
1.3   Sources of Extension Funding by Year.................... 15

2.1   Open Systems Model 28
2.2   Three-Stage Model of Program Evaluation 30
2.3   Systems Effectiveness Model for the Cooperative Extension Service 33

3.1   Awareness of Extension and Its Programs 48
3.2   Percent Aware of Extension or Its Programs by Region.. 49

4.1   Household Use of Extension by Region.................... 61
4.2   Household Use of Four Program Areas, 1981 67

5.1   Satisfaction with Extension by Region of the U.S. 76
5.2   Satisfaction with Extension Programs by Users of the Service..................................... 77

6.1   Support for Extension by Users of the Service 92
6.2   Support for Extension Programs 93
6.3   User Support by Region of the U.S........................ 94

7.1   County-Based Effectiveness Model 103
7.2   Findings for County-Based Effectiveness Model 111

8.1   Systems Effectiveness Model Reexamined 116

A.1   National Survey Instrument 164

B.1   Standardized Correlation Coefficients for County-Based Effectiveness Model 188

Acknowledgments

Our ideas and perceptions of reality continually evolve. Many friends, colleagues, and scholars have contributed to the generation of the ideas in this book, and it would be impossible to acknowledge all of their contributions. The references we have cited name many who influenced our approach to evaluation and our understanding of the Cooperative Extension Service, but that list cannot indicate the numerous intangible contributions made by others.

The desire to assess the scope and impact of Extension began for Paul Warner during his work with Howard Phillips, Clarence Cunningham, Dick Young, and others while at The Ohio State University. Jim Christenson has greatly relied upon his exposure to Extension through Vance Hamilton, Tom Hobgood, John Collins, Paul Thompson, and Mauri Voland while working as an Extension Specialist at North Carolina State University.

The direct impetus for this book grew out of earlier projects of the Kentucky Cooperative Extension Service and the Kentucky Agricultural Experiment Station that focused on citizens' needs and community services. Support for the development of the concepts and eventually for the book was provided under an Extension Service U.S. Department of Agriculture special project grant entitled "Use of General Population Surveys as a Method for Assessing the Impact of Extension Programs." Fred Woods served as our liaison on the project and assisted in the process of utilization within the federal structure. Claude Bennett and his colleagues with the PDEMS staff, Extension Service, USDA, provided important encouragement and suggestions. Support for conducting the national survey was provided by a grant from the Ford Foundation to co-workers Larry Busch and Bill Lacy through the Committee on Agricultural Research and Extension Policy at the University of Kentucky. To both Larry and Bill, we owe a sincere debt of appreciation. Special acknowledgment goes to David Brown, Associate Director of the Economic Research Service, USDA, for his encouragement, support, and assistance in making this book a part of the Rural Studies Series of the Rural

Sociological Society, and to Lynn Arts of Westview Press, who guided the publication process.

Several colleagues and friends read early drafts of this manuscript, providing very helpful suggestions. We are especially grateful to Don Dillman and Ronald Powers for their willingness to provide ongoing feedback over a period of almost a year. We also wish to thank Jim Hildreth, Clarence Cunningham, Cindy Noble, Dan Moore, Sara Steele, C. Milton Coughenour, Richard Maurer, and Lori Garkovich for commenting on various parts of the book.

We are also appreciative of the patience and dedication of Nona Richie, who typed and retyped most of the manuscript. Rosemary Cheek, Michael Claycomb, and Lynn Webb assisted in typing and illustrating the material. Ann Stockham helped to edit and index the manuscript. We are grateful for the help of Janice Taylor, who skillfully expedited the administrative details associated with the grants and book. Initial presentations at the 1983 Rural Sociological Society meetings, at Washington State University in late 1982—facilitated by Bob Butler—and at the American Association of Adult and Continuing Education meetings organized by Joan Thomson helped to provide valuable feedback during the final stages of preparing the manuscript.

Finally, we would like to extend our appreciation to Charles E. Barnhart, Dean of the College of Agriculture; Shirley Phillips, Associate Director of the Kentucky Cooperative Extension Service; and C. Oran Little, Associate Director of the Kentucky Agricultural Experiment Station, for their feedback and encouragement. Obviously, the interpretations of findings and the policy recommendations found herein are our own, and we take full responsibility for them. Finally, we would like to thank our wives, Ruth and Patti, and our children, Kevin, Tim, Eric, and Kellie, for their support and love.

Paul D. Warner
James A. Christenson

Introduction

What ought to be the role of Extension in the information society of the 21st century? Should Extension's mission be broad-based or narrowly defined? Should staff be primarily generalists or specialists? What is Extension's public image? Should Extension's clientele be primarily rural or urban, farm or nonfarm? Should programs emphasize personal or impersonal communication methods? And who will champion Extension's cause in securing future support?

The Cooperative Extension Service (CES) as a publicly supported educational agency is continually struggling to define its proper function and purpose in a changing society. Issues of defining appropriate target audiences, delivering quality programs in the most efficient manner, projecting a positive organizational image, and maintaining an adequate support base are being widely discussed. Some observers contend that changes have been too slow in coming, that the organization has not been responsive enough to the needs of the people; while others are critical of Extension for going too far and see the organization as straying from its original purpose. This book addresses the dilemmas facing Extension as it defines its role as an educational agency.

This study provides the first comprehensive nationwide public assessment of the Cooperative Extension Service. There have been other studies that have examined specific program areas such as 4-H, home economics, and agriculture; but none has focused on the total organization. This study is a public's assessment of an agency's programs. The primary source of information is a telephone survey of a national sample of the general population. The respondents are people selected at random; they are not from agency clientele lists. Inasmuch as agency evaluations are frequently criticized or ignored because they are seen as biased expressions of selected clientele by agency personnel, this study was conducted independently of the organization. In addition to the national survey, supporting evidence was also drawn from similarly conducted state studies.

Most observers would agree that, over the years, Extension's programs have grown substantially with few systematic efforts to provide for comprehensive, in-depth evaluations. Traditionally, evaluation has been concerned with measuring inputs in terms of the methods used to reach people (i.e., meetings, office visits), counting the number of participants, and receiving informal feedback as to the happiness of clientele with specific activities and events. The primary vehicle for recording such information has been the Extension Management Information System (EMIS). Studies that have gone beyond these basic counting procedures often have been carried out in limited geographical areas (i.e., one county or area) or of specific programs (i.e., Expanded Food and Nutrition Education Program), or both. They have provided interesting case studies and anecdotal materials, but it is unclear what value they have for generalizing to other situations.

Most of the concern for the evaluation of public agencies has come from decision makers concerned with broad policy issues, administrators with responsibilities for program implementation and resource allocation, and evaluators interested in improving the efficiency of program delivery. Even then, after their questions have been answered, the issues identified may bear little or no resemblance to those that are important to clientele and the public. As Katz et al. (1977:1) assert, "There is a vast and profound neglect of the perceptions, experiences, and reactions of the people who themselves are supposedly being served." This lack of public reaction to so-called public services is a serious shortcoming. Effective mechanisms for registering client feedback are especially critical with public agencies in that there are no alternative sources of services to which clientele can turn. There is no market place as in the private sector. Though evaluation of public agencies from the standpoint of clientele is widely supported, there are few such studies available.

Why a book on the evaluation of Extension? In recent years, there has been increased emphasis on evaluating the programs of the Cooperative Extension Service for purposes of accountability and for gathering information in order to improve such programs. In fact, it seems like all of a sudden everyone is interested in evaluating Extension. An evaluation of Extension was mandated in the 1977 Food and Agriculture Act. This stimulated agency personnel to give increased attention to the importance of evaluation. At about the same time the Extension Committee on Organization and Policy (ECOP) appointed a task force that was charged with the preparation of a position statement on evaluating Extension programs. Soon thereafter (1981), the General Accounting Office (GAO) released a study that questioned the mission of Extension. And then in 1982, the Subcommittee on Department Operations, Research, and Foreign Agriculture of the House Agriculture Committee held oversight hearings on Extension. Similarly, there have been questions raised about Extension's role and effectiveness in individual states. But no comprehensive national assessments have been conducted until now.

Again, why a book on the subject? Why not publish a report, a bulletin, or merely a series of news releases? After all, those are Extension's primary delivery methods. Well, one of the most striking aspects of the body of literature on Extension is its absence. If you go to the agricultural library of a land-grant university, you will likely find a number of classics on Extension philosophy and methods written in the 1920s and 1930s. Since that time, most of what is available could be referred to as fugitive literature. It is in the form of mimeos, research reports, theses, and papers often available only from the author. Therefore, there is a need for more permanent additions to the body of knowledge on Extension. In addition, this study contributes to the emerging area of evaluative research in that a comprehensive model for evaluating a public service agency is developed and illustrated.

Questions addressed in the book are: What is the extent of public awareness of Extension? Who really uses its services? Are they primarily rural or urban, rich or poor, young or old? How do they assess the quality of the services? And, then, what level of funding support do they desire for the organization in the future? In addition, county-level information on program inputs and operations are related to the clientele's assessment. An organizational model is presented that serves to guide the reader through this series of questions.

This book is intended for persons who are making decisions about the future of Extension and its programs: decisions that are made by such persons as county staff, program development specialists, evaluators, administrators, legislators, local government officials, and citizens' groups. It is also our hope that it will be used in the training of future Extension staff, both at the undergraduate and graduate levels. In addition, this effort will contribute to the emerging field of evaluative research.

This book is "plowing new ground" in that it contains many unique features. First, it is the first national survey that looks at the total Extension organization rather than just one program area. Second, it provides an expression from the viewpoint of the environmental actors, the clients and the public, rather than the organizational staff. And third, it combines survey information with information recorded by the organization (through EMIS and other materials). This approach not only makes use of survey methods but also increases the utility of traditional reporting systems. It is then possible to examine the awareness and utilization of the agency's services, to describe the characteristics of the clientele, to determine whether users are satisfied with the service and are willing to continue to see it supported with public monies. In addition, these expressions by clientele will be related to such programmatic elements as staff time expended, cost of the program, and educational methods used. Together these elements are developed into a comprehensive evaluation model of the Cooperative Extension Service. We have organized the book in the following manner:

In Chapter 1, we review the historical development of the Cooperative Extension Service as a public agency. The purpose of this account is not to provide a comprehensive historical picture but, rather, to identify a number of critical elements that have influenced the development of Extension over time and to highlight dilemmas facing Extension in the future.

In the second chapter, we develop the overall framework for the evaluation study. This chapter presents the Systems Effectiveness Model, discusses the major organizational elements, and outlines their relationships.

Chapter 3 begins the discussion of specific aspects of the model. It answers the questions: What is the extent of public awareness of Extension in general and specifically for the four program areas of agriculture, home economics, 4-H youth, and community development?

The clientele of Extension are identified in Chapter 4. Included in the discussion are the proportion of the population that uses the services of Extension, the program areas they utilize, a description of characteristics of users and nonusers, and a test of equal opportunity criteria.

Chapter 5 focuses on the level of satisfaction of clientele with Extension programming. It not only provides an indication of overall satisfaction levels, but it also serves to identify the characteristics of those who are satisfied and dissatisfied.

The level of support desired for Extension is identified in Chapter 6. This measure provides further indication of the value or worth associated with this agency.

Chapter 7 focuses on county programs. It differs from the others in that it relates county data collected by the agency through EMIS and other methods to public survey information. The emphasis in this chapter is on the relationship between educational methods and client response.

Finally, in the concluding two chapters the overall findings are summarized, future priorities outlined, and policy issues delineated. We have tried to avoid restating the obvious and have concentrated on new insights from this approach.

1
Extension in Changing Times

> *The Cooperative Extension Service is a unique achievement in American education. It is an agency for change, a catalyst for individual and group action with a history of nearly 70 years of public service.*
> —USDA-NASULGC, "Extension in the '80s Report," 1983:1

What will Extension look like in the year 2000? Can an organization created in 1914 make the adjustments necessary to survive the rapid and pervasive changes occurring in American society? We are entering a new era unlike any we have known before. Futurists predict that the changes that the organization has experienced in the past are nothing compared with what it is likely to face in the future. As we move into an era when information is increasingly important, Extension, as an educational agency, can play a pivotal role. Change is inevitable, the only question is Extension's response. Extension has the opportunity to shape its future, or it can react to a future shaped by others.

Extension is unique in structure and function. It has not been restricted to a single program or activity but, rather, has been allowed to adjust to changing needs. Society has allowed few organizations this flexibility. Organizations tend to be created for a single purpose. Extension has defined its role as education, even though the subject matter and audiences have changed over time. Furthermore, Extension has not been captured by any one level of government. All three levels (federal, state, and local) share in its support and control.

In times of limited resources, there is increased pressure on public agencies to demonstrate their worth to society. Extension's mission is being questioned and new priorities are being established. Extension is being forced to make decisions it has never had to make before, many of which are difficult and unpopular. With a budget of $800 million and a staff of 17,000, it is not surprising that Extension is being scrutinized.

Historically, Extension has focused on program development and implementation. The assumption has been that if programs are carried out and people are reached, then the program is effective. However, today such assumptions are unacceptable. Just because a large number of people are exposed to a program does not necessarily mean that it has an impact on their lives. A desire to identify Extension's contribution to the public good, coupled with the current period of budget tightening, have led to a greater emphasis on Extension evaluation.

Recent agency reports and external reviews have called for an increased emphasis on evaluation. There is the realization that evaluations can serve many functions. They can document impacts, provide information on alternative programming methods, document compliance with programming guidelines, identify resources utilized for specific programs, provide an accounting of the relationship between benefits received and resources used, and indicate whether program objectives have been met. In reality, past evaluation studies have done little of the above. They have tended to focus on information generated by agency staff, they have been limited to a specific program or geographical area, and they have been largely descriptive. What is needed is a more comprehensive picture of Extension's performance. Is the organization as a whole judged as effective? Is it accomplishing what it is supposed to do?

This book examines the Extension Service as a total organization; not as separate programs and activities, but as a whole. This study is also national in scope. It is truly the first comprehensive nationwide assessment of Extension. This effort relates the resources that go into the agency, the programs that are implemented, and the ultimate impacts on the public. It identifies Extension's image, it examines who Extension is serving, it determines whether clientele are satisfied with the service, and it assesses future support for the agency.

Now, let's review the origins of Extension, its legislative mandate and organizational structure.

Extension's Roots Go Deep

Extension is a national resource. For nearly three-quarters of a century, the Cooperative Extension Service has been an important influence on the development of rural America. Extension was created by the Smith-Lever Act in 1914 as a third arm of the land-grant system in order to transmit information from colleges and the Department of Agriculture to local people. According to the purposes specified in the original legislation, Extension is to disseminate and encourage the application of useful and practical information relating to agriculture, home economics, and related subjects among the people of the United States not enrolled in land-grant colleges.

Extension had its beginnings in the movement to improve agricultural production. Seaman Knapp's successful use of the demonstration method on Louisiana farms in the fight against the cotton boll weevil served as the model for the legislation. The Smith-Lever Act uses the terms cooperative agricultural Extension work, emphasizing the farm thrust of the legislation. Until the 1930s, Extension was seen as the primary agency representing the United States Department of Agriculture in local communities. However, with the thirties came a multitude of public agencies to cope with the impact of the depression upon agriculture and rural areas. The Agricultural Adjustment Act served to stabilize farm prices; the Farm Security Administration expanded farm credit services; rural homes and farmsteads received electricity as a result of the efforts of the Rural Electrification Administration; and the preservation of land resources was promoted by the Soil Conservation Service. These agencies came into being to provide specific services to farm and rural residents and were established at the local level largely through the organizational efforts of Extension (Ballew et al., 1976).

The 1940s and 1950s began a period of rapid technological advancement in American agriculture. Farmers improved their competitive position primarily through the adoption of improved management practices, the expansion of their resource base, and increased efficiency. This led to production surpluses and a heavy reliance upon agricultural production in international trade and development assistance. With an abundance of reasonably priced food supply pretty much assured, national attention, and to some extent the concerns of Extension, in the 1960s shifted to the problems of urban residents, low-income persons, and minority groups.

At the same time, there was a substantial reduction in the size of the farm and rural population. In 1910, approximately 35 percent of the total United States population were classified as farmers (Figure 1.1); and, until 1940, almost one-quarter of the population were farmers. Since that time, the farm population has gradually declined to less than three percent of the total population. A similar decline occurred in the rural portion of the population. At the turn of the century, over 50 percent of the nation's population were defined as rural; while in 1980, less than 30 percent were labeled rural. As these rural and farm residents relocated in metropolitan centers, they carried with them an awareness of what Extension did for them in rural areas, thus, setting the stage for the beginning of an urban clientele.

The identification of home economics as appropriate subject matter for Extension was clearly spelled out in the Smith-Lever Act. In fact,

FIGURE 1.1
Farm and Rural Population, 1910 to 1980

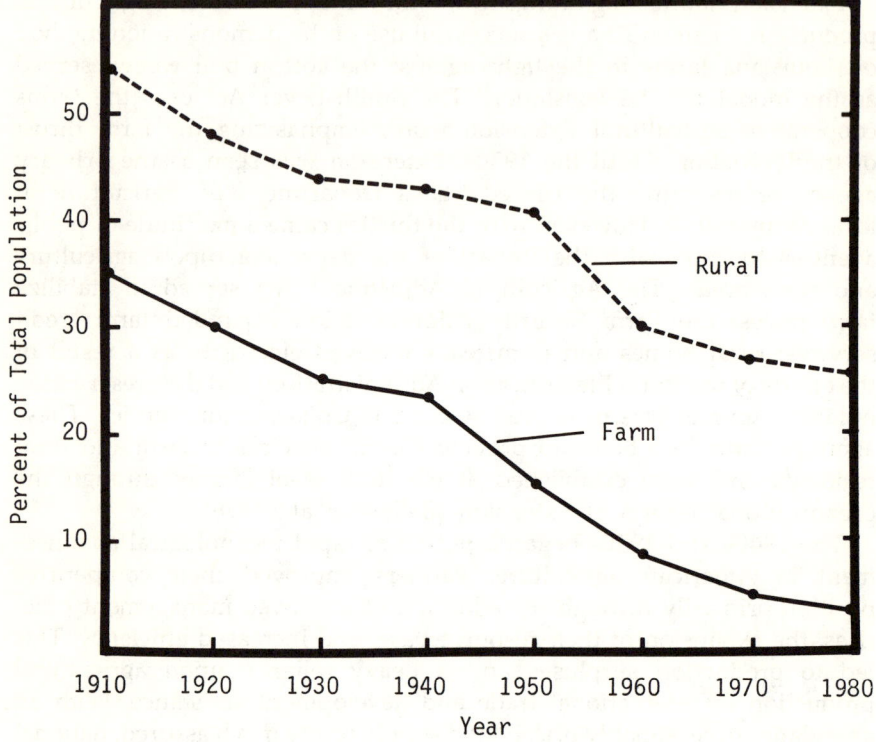

Source: Information taken from reports of the Census of Population and Census of Agriculture. (See Chapter 1, Note 1 for citations and definitions.)

in testimony before the House Agriculture Committee concerning the inclusion of home economics in the bill, Congressman Lever stated that the "committee believes there is no more important work in the country than this [home economics]" (United States Congress, 1913:6). The testimony also made it very clear that young people as well as adults were to be included in this educational program. In the same hearings, Representative Lever indicated that "one of the main features of this bill is that it is so flexible as to provide for the inauguration of a system of itinerant teaching for boys and girls" (p. 5-6). Another important aspect of the legislation is its reference to the problems of not only the farm and the home but also the rural community. Therefore, instead of being narrowly defined, the mandate of the Smith-Lever Act was actually very broad and included such goals as "better living, more happiness, more education, and better citizenship" (United States Congress, 1913:5).

Flexibility has been one of Extension's strengths. "In its early history, Cooperative Extension's role appeared straightforward and limited. Extension was the organization best equipped to attack informal educational problems of production agriculture and rural living. There was a close interrelationship of farmers and rural residents with Extension" (Hildreth and Armbruster, 1981:853). However, with changes in the nature and scope of the rural farm population, and society in general, have come new mandates through the legislative process. At least some of the broadening of the clientele base can be explained by programs developed in response to federal directives. Most earmarked funds have been directed toward nontraditional audiences—especially low-income, urban residents. Recent legislation has expanded Extension's role to include such topics as nutrition education, gardening, energy, and rural development.

At the same time, Extension functions through locally based offices that attempt to be responsive to a variety of clientele with differing needs. In agriculture, for example, services are extended to large commercial farmers, to limited resource farmers, and to backyard gardeners. In home economics, traditional homemakers, women employed outside of the home, and single parents are served. Youth programs serve teen leaders, farm boys and girls, and urban children; while community development lends assistance to local officials, low-income community groups, and tourist agencies. With these varied programs and clientele has come a diversification of subject matter.

Over the years, many new programs have been added and new clientele reached, often without eliminating others. For example, in agriculture, there is a smaller number of farmers today, but food and fiber are as crucial to the nation and world as ever before. Some would argue that, with the specialization and concentration of our present production and distribution system, agriculture becomes more vulnerable and thus relatively more important. County Extension staff must be acquainted with more subject matter and with greater levels of sophistication. The agricultural agent is expected to be knowledgeable on such diverse topics as the application of pesticides, contracting for the sale of commodities, identifying insect damage on alfalfa, and balancing a least-cost animal ration. In addition, the same agent is now serving an urban audience with information on such topics as leaf composting, master gardening, and landscape horticulture. Similar changes have occurred with 4-H and home economics programs. Generally speaking, Extension has broadened its role with existing clientele and at the same time has acquired new audiences.

TABLE 1.1
Distribution of Client Contacts by Program Area, 1982

Program	Client Contacts	Percent
Agriculture	34,502,480	32
Home Economics	28,350,304	27
4-H/Youth	37,627,523	35
Community Development	6,554,226	6
TOTAL	108,664,700	100

Source: Extension Service, U.S. Department of Agriculture, Annual Report, 1983.

The variety of program contacts is demonstrated in Table 1.1. The greatest number of contacts with clientele is reported in the 4-H program (35 percent). It is followed by agriculture and home economics (about 30 percent each) and then drops down to 6 percent with community development.

A Unique Organization

Extension programs are educational in nature and, therefore, are appropriately placed in an educational system (e.g., the land-grant colleges). Unlike most other agencies, the CES does not have financial or regulatory powers in the implementation of specific programs, nor does it conduct formal classroom instruction. It provides informal, noncredit education for the purpose of assisting individuals in making their own decisions.

Extension prides itself in its responsiveness to local needs and priorities. As a voluntary educational institution, programs must appeal to local needs in order for Extension to maintain clientele. Though a strong critic of Extension, after visits to a number of counties, Susan DeMarco concedes that, "Probably the major strength that Extension has going for it is the perception, at least among its own clients, that while government is something out there somewhere, Extension is local and responsive" (1980:3).

Except for about 100 professional staff at the federal level, Extension staff are employees of the land-grant institutions and are not direct-

TABLE 1.2
Distribution of Staff by Type of Position, 1982

Type of Position	Number	Percent
Federal	116	1
State Staff		
Directors and Administrators	507	3
Subject Matter Specialists	3,706	22
County and Area Staff		
County Agents	11,240	67
Area Agents	629	3
Program Leaders and Supervisors	651	4
TOTAL	16,849	100

Source: Explanatory Notes, Science and Agriculture Administration 1984 Budget, 1983.

line employees of any of the three levels of government. From time-to-time, there have been attempts by one of the partners to treat Extension as an administrative organization for implementing policy. However, its unique administrative arrangement has served to preserve the educational nature of the program and has reduced the amount of political influence on the institution.

Fifty state and four territorial Extension Services deliver programs at the local level and, correspondingly, most of the staff are officed within the 3,150 counties across the nation. State and area specialists, in turn, provide backup support to the county staff. In addition, paraprofessionals are often employed for specialized programs. The emphasis on county programming is reflected in the relative number at each level (Table 1.2), with two-thirds employed as county staff.

TABLE 1.3
Distribution of Staff Time by Program Area, 1982

Program	Staff Years	Percent
Agriculture	7,975	36
Home Economics	7,166	32
4-H/Youth	5,856	26
Community Development	1,414	6
TOTAL	23,348	100

Source: Extension Service, U.S. Department of Agriculture, Annual Report, 1983.

Most states then organize their staff around four program areas: agriculture, home economics, 4-H youth, and community development. Table 1.3 shows the national distribution of staff by the four program areas. Though the largest number of staff is devoted to programs in agriculture, it is followed closely by home economics and 4-H. The 4-H program utilizes 26 percent of the staff but makes 35 percent of the total contacts. There has been a slight increase in the total number of staff over the past three decades (Table 1.4), but the distribution has remained relatively unchanged. Over the past 25 years, personnel numbers have increased less than one percent per year.

Though the basic unit of Extension programming is at the county level, county staff frequently seek support from state and area specialists. With increased specialization has come the need for more detailed technical expertise, sometimes beyond what the county staff can provide. Fortunately, county staff are tied into the land-grant system so the expertise can be found within the Extension organization. That makes the staff the local linkage to the technical information system. This change in role for the county staff led DeMarco (1980:23) to conclude that "The county staff may be becoming more important to the process of information transfer and education than they are to the actual content."

The number of specialists increased substantially during the fifties and sixties to the extent that now there is one specialist for every three county staff. As can be seen in Figure 1.2, that ratio varies from almost a one to one ratio in agriculture and community development to six

TABLE 1.4:
Number of Extension Personnel By Type, 1914-1982

Year	Directors and Administrators	State Specialists	Leaders and Supervisors	Area Agents[a]	County Agents	Total
1914	50	221	112	0	1,237	1,620
1918	115	512	575	0	5,526	6,728
1928	106	1,004	376	0	3,675	5,161
1938	131	1,551	493	0	6,507	8,682
1948	159	1,933	596	0	8,785	11,473
1958	217	2,554	754	0	11,124	14,649
1968	295	3,850	695	0	10,220	15,060
1978	487	3,410	696	732	11,342	16,667
1982	507	3,706	651	629	11,240	16,733

Source: U.S. Department of Agriculture, 1980a:30 and Explanatory Notes, Science Agriculture Administration 1984 Budget, 1983.

[a]The category of Area Agent was not used prior to 1969.

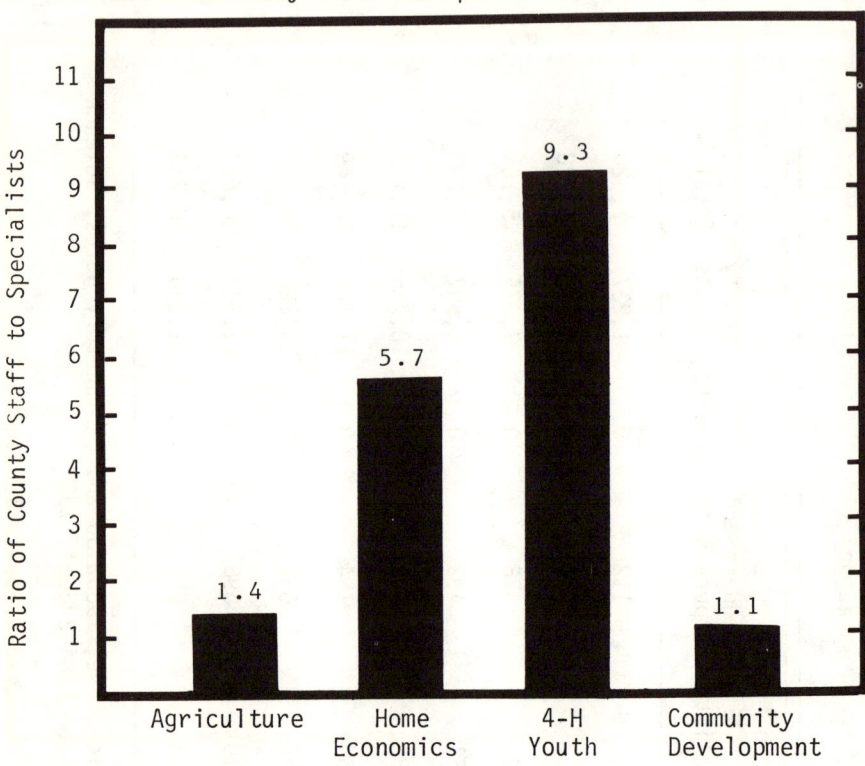

FIGURE 1.2
Ratio of County Staff to Specialists

Source: Compiled by Personnel Devision, Extension Service, U.S. Department of Agriculture.

county staff for each specialist in home economics and a nine to one ratio in 4-H.

Extension programs are also highly dependent upon volunteers. Each year the 4-H and home economics programs alone involve over a million volunteer leaders. This volunteer effort is equivalent to 86,000 staff-years, or more than five times that of the organization's professional staff time. The 4-H and home economics programs continue to depend upon an organized club approach for carrying out their programs; though, in some states, the club structure is a minor portion of their programs or has disappeared entirely.

An Unprecedented Funding Arrangement

The Extension Service is truly a "cooperative" venture among three levels of government—federal, state, and local. All three share in the

FIGURE 1.3
Sources of Extension Funding by Year

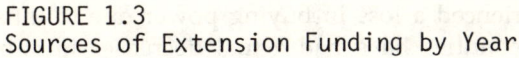

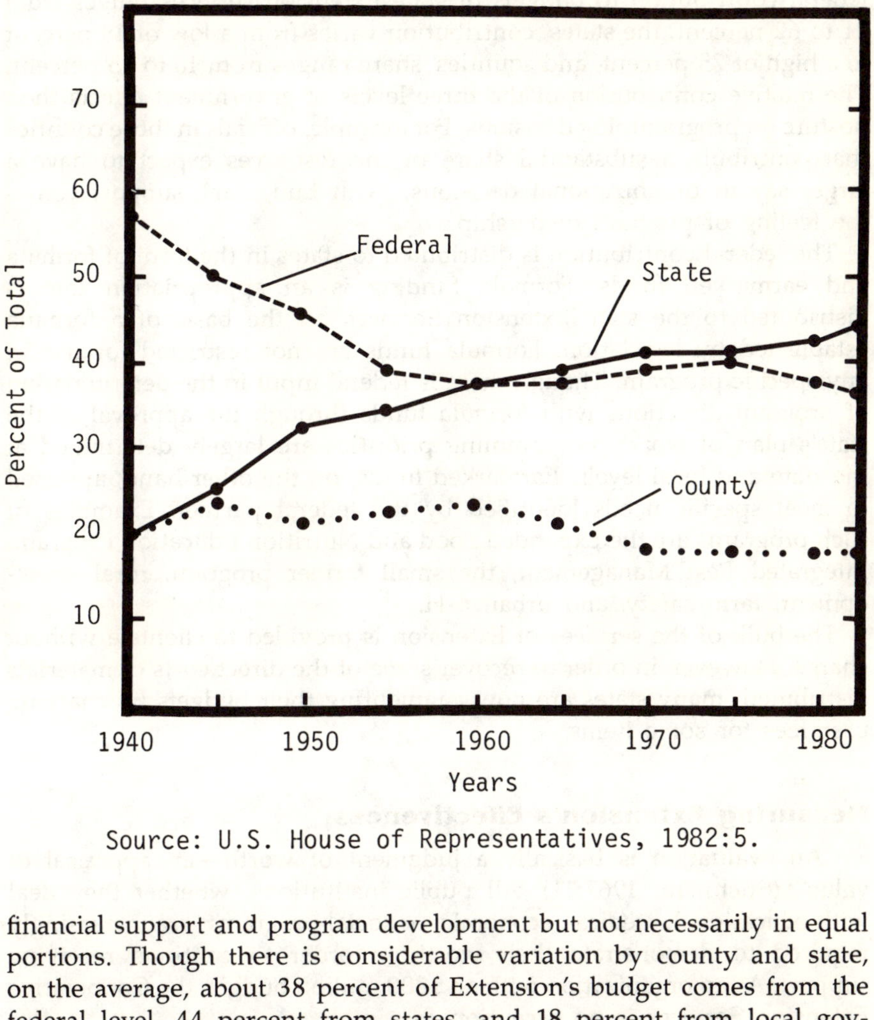

Source: U.S. House of Representatives, 1982:5.

financial support and program development but not necessarily in equal portions. Though there is considerable variation by county and state, on the average, about 38 percent of Extension's budget comes from the federal level, 44 percent from states, and 18 percent from local governments (1983 figures from Explanatory Notes, Science and Agriculture Administration 1984 Budget, 1983). This balance in funding support has been quite stable since about 1955 (Figure 1.3). However, since 1975, the federal portion has begun to decline, while the states' portion has increased. At this time it is impossible to determine whether this is the beginning of a trend or is merely a short-term fluctuation.

This shift in funding support from the federal to the state level is based on overall averages and is not necessarily reflective of individual states. Not all states have increased their support. Many state Extension

Services have experienced a loss in buying power due to inflation, and certain states and counties have had actual decreases in support. The federal contribution to budgets of state Extension services ranges from 21 to 62 percent, the states' contribution varies from a low of 18 percent to a high of 75 percent, and counties' share ranges from 10 to 45 percent. The relative contribution of the three levels of government affects their posture on programming decisions. For example, officials in those counties that contribute a substantial share of the resources expect to have a larger say in organizational decisions. With budgetary support comes the feeling of program ownership.

The federal contribution is distributed to states in the form of formula and earmarked funds. Formula funding is an appropriation that is distributed to the state Extension Services on the basis of a formula established by legislation. Formula funds are not restricted for use in any specific program. Though there is federal input in the determination of program directions with formula funds through the approval of the state's plan of work, programming priorities are largely determined at the state and local levels. Earmarked funds, on the other hand, are used to meet specific needs identified by the federal partner. Examples of such programs are the Expanded Food and Nutrition Education Program, Integrated Pest Management, the small farmer program, rural development, farm safety, and urban 4-H.

The bulk of the services of Extension is provided to clientele without charge. However, in order to recover some of the direct costs of materials distributed, many states are now augmenting their budgets by charging user fees for some items.

Measuring Extension's Effectiveness

"An evaluation is basically a judgment of worth—an appraisal of value" (Suchman, 1967:11). All public institutions, whether they deal with educational, medical, economic, or social programs, are increasingly required to demonstrate their worth in order to continue receiving support. As an organization with an $800 million budget, the Cooperative Extension Services is no exception.

Most state Extension services have utilized a planning-by-objectives approach to program planning and reporting. Odiorne (1965) developed this approach as a complete system of management, planning, and control. This goal-oriented system has been seen as providing the necessary information for purposes of accountability. It contains a plan of work that includes planned effort for the year and a narrative report at the end of the period. In addition, information feeds into the Extension Management Information System (EMIS). Over the years this reporting system has undergone changes, but it continues to be utilized pretty much as it was originally envisioned. It provides a reporting of county and state data on program activities and clientele served. Such a system provides information for monitoring the performance of staff and for assessing the scope of the program. This approach is goal directed in that program objectives are established at the beginning of a planning period and are reported against during program implementation. This planning and reporting system provides important information for program evaluation; however, it is only a tool for collecting data. It is not an evaluation in itself. It was probably seen as adequate when it was established, though today it is viewed as only one part of a more comprehensive evaluation system.

The 1977 Food and Agriculture Act mandated an overall evaluation of Extension; the results of which were released in 1980. Specifically, the Secretary of Agriculture was instructed to conduct "an evaluation of the economic and social consequences of the programs of the Extension Service and Cooperative Extension Services, including those programs related to agricultural production and distribution, home economics, nutrition education, community development, and 4-H youth programs" (Food and Agriculture Act of 1977, Section 1459, Title XIV). Conducting a national evaluation of Extension that focuses on program impacts is a difficult task due to the lack of consensus as to appropriate measures of consequences and the diversity of programs in terms of subject matter and geographical areas. As a result, the report is short on "impacts" and long on documenting participation and activity levels of programs. Likewise, there is substantial variation by program area. Though these types of efforts tend to be constructed from agency records and are often viewed with suspicion by those outside of the agency, it does serve to focus attention upon the need for increased emphasis on evaluation.

Such was the case in the appointment of a national Extension Committee on Organization and Policy (ECOP) Task Force on Evaluation. The Task Force called for the establishment of a national system of evaluation that would include three types of information from: (1) studies of selected high priority programs, (2) system-wide accomplishment data on a sampling basis, and (3) comprehensive information on participants served and resources utilized (Task Force Report:1981). In addition, it recommended the establishment of a national accountability and evaluation staff unit to give leadership to efforts in the states and the implementation of a four-year planning and reporting cycle.

Besides the recommendations for national evaluation efforts, the Task Force called for more evaluation to take place at the state and county levels and for these approaches and findings to be shared with other states. The specific activities being conducted in individual states and counties are numerous and varied. Summers et al. (1981) summarized these efforts in a report entitled "Program Evalation in Extension."

In 1981 the General Accounting Office (GAO) released a study questioning the lack of a clearly defined mission statement for Extension. This deficiency was seen as especially critical in the present atmosphere of increased demands and fiscal constraints. The GAO report related the mission to the funding and structure of Extension in what could be seen as a "catch 22" situation. It concluded,

> If the Extension Service is to be a socially oriented organization with broad educational and behavioral modification objectives, then changes may have to be made to its basic funding formulas and organizational structure. On the other hand, if its mission is to be limited to more traditional focuses, then the scope of its programming may have to be reduced (1981:21).

In other words, if Extension is to provide broad educational programs to clientele other than farmers, then it does not belong in the U.S. Department of Agriculture, nor should it be funded according to a formula largely dependent upon the rural and farm populations. However, if it chooses to restrict its programs to farmers, then the present level of resources cannot be justified based on the small size of the farm population. Either way Extension would experience drastic changes. The report also called for an increased role of the federal partner in Extension program development and an examination of its methods of evaluating and accounting procedures.

Following the GAO report, the Sub-committee on Department Operations Research and Foreign Agriculture of the United States House of Representatives Committee on Agriculture held oversight hearings on the operations of Extension. The objective of the oversight hearings was to "establish, from the Congressional viewpoint, the broad boundaries of an organizational mission for the CES in the last decades of the twentieth century" (U.S. House of Representatives, 1981:2).

The national evaluation, the ECOP Task Force, and the GAO report all cite the need for improved evaluation and accountability in Extension. They suggest the desirability of identifying agreed upon performance and impact measures that are, to the extent possible, uniform across the nation. State and program-specific indicators could then serve to support these common elements. Other than compiling information on program participation and allocation of staff time, few Extension evaluation studies have been system-wide. In fact, almost all evaluation studies have been limited to a single program area (Kappa Systems, 1979). And as Bennett

(1982:164) concludes, "Extension's overall accountability efforts, then, are reduced as a rule to reports on the major Extension program areas." This focus on the individual program area aids decision making internally but does not speak to the broader organizational needs. Because the scope of the agency is broad and the subject matter diverse, it is difficult to establish clear criteria for identifying clientele and assessing overall organizational effectiveness. Yet, the future of Extension depends on its ability to document its impact, to demonstrate its effectiveness, and to maintain a positive political climate for the organization as a whole.

Establishing Legitimacy

It is generally assumed that an organization has a protected status as long as its products are considered important to society (Perrow, 1970). This legitimization is derived from the organizational environment—from such groups as clientele, interested citizens, and taxpayers. They find the outputs of the organization desirable and, therefore, want the organization to continue. Legitimization occurs because the product of one part of society is needed by other parts, thus, attributing value to the output and the organization producing it.

In the private sector, legitimization is generally thought to be derived from the marketplace. Individuals express their support by purchasing its products. Public agencies, in contrast, derive their legitimacy from their social and political environments. Public organizations can be seen "as competing 'politically' for 'institutional legitimacy' in contemporary environments and less as competing 'technically' for 'market advantages'" (Meyer and Associates, 1978:365). In addition, the product of a public agency, especially one like Extension that deals with education, may be difficult to identify, thus, leaving legitimization to social agreement and political processes.

The legitimacy of Extension lies in the uniqueness of its informal educational program to help people develop their own potentials. While the clientele of Extension has broadened substantially over the years, there is still a strong tendency to define Extension's base of support as farmers, agricultural commodity groups, homemakers, and rural people. With this limited perspective and with a decline in the number and impact of farm residents, the legitimacy of Extension increasingly comes into question. Though agriculture is as important today as ever before, some Extension observers conclude that the organizational resources should decline with the number of farmers.

There is a tendency among organizational managers to assume that once legitimacy is conferred, it will always be present; when, in reality, the relationship of the organization with its environment is ever changing.

As a result, organizations find, much to their surprise, that the usefulness of their goods and services is questioned; and individuals are seeking to restrict their resources or to withdraw their "protected status" (Perrow, 1970). Organizational managers often become preoccupied with matters of internal performance and take the legitimacy of the organization for granted. They then find themselves facing a crisis of legitimacy in that the very existence of the organization is in jeopardy. Whereupon, they rally all of the resources available in an effort to "save the organization." And once the crisis has passed, they again return to focusing upon internal matters and ignore the bases of legitimacy.

In order to establish and maintain organizational legitimacy, administrators must recognize the grounds of legitimacy, with whom it lies, and the possibility that it is likely to shift over time. In the long run, legitimacy cannot be handled merely as a response to crisis; but it must be a continuous process of communication between the organization and its environment—including not only traditional clientele groups but also funding bodies, political decision makers, and the general public.

Organizational Stereotypes: Myths or Truths?

Over the years there have developed a number of widely held generalizations about Extension. Some have grown directly out of the philosophy of the organization, while others have evolved out of a long history of how Extension has carried out its educational programs. Some are no doubt true, while others are merely myths. Of concern is whether there is congruence between reality and the perception of reality (Warner and Christenson, 1983). This study will examine some of these assumptions and, to the extent possible, suggest which can be substantiated.

"Everyone Has Heard of the Cooperative Extension Service"

It has generally been assumed that Extension has a high level of organizational visibility, especially with its traditional audiences. This would be expected to be the situation, especially in rural areas, inasmuch as Extension has a long history of work with farm and rural people. However, what is the extent of its identity among urban residents? Do they know that Extension even exists? Or is Extension lost among the myriad of government agencies?

It is very possible, because of the diversity of its programs, that instead of a single identity Extension actually has a pluralistic identity. For different clientele, Extension may be known as 4-H, homemakers clubs, development organizations, or agricultural commodity groups. This problem would likely be amplified among nontraditional clientele. When county Extension staff identify themselves as 4-H agents or agricultural agents, it is not always clear they work for the same organization. It is possible there is more name recognition for the programs of Extension than for the organization itself. Extension has not felt it necessary, nor desirable, to promote itself as an organization. Instead, it has operated on the philosophy that a job well done is the only advertisement necessary. In fact, the very process upon which Extension programming is based encourages staff to work in such a way as to make clientele feel that the idea is their own. Extension has not been concerned with getting its share of the credit for the contribution it makes. However, in the present era of competition for limited resources, more attention may have to be given to Extension's identity.

"Farmers and Rural People Are the Clientele of Extension"

Extension was established to disseminate the findings of researchers at land-grant universities to persons not resident in institutions of higher education. This suggests that Extension is in the business of communicating ideas, practices, and technologies to whomever is interested in using them. As a result, participation in this educational program is voluntary in nature. Thus, the clientele are, to great extent, determined by self selection. As Jenkins (1980) concludes, because America is largely middle class, so are Extension clientele.

Extension has strived to provide clientele with the programs that they desire. This has occurred through a system of involvement of citizens in the program-planning process. However, since its beginning, Extension has been careful to try to remain an educational institution, not a service agency.

As an outreach arm of the land-grant universities, it is generally concluded that Extension has been able to reach large numbers of people. Is this true? And what proportion of the population does this represent? What proportion should it be expected to serve? Traditionally, Extension's clientele have been referred to as middle-class, rural residents. But who are Extension's clientele today? To what extent has there been expansion into new geographical areas in response to changing societal needs? Does the present mix of clientele include adequate numbers of such groups as low-income individuals, minorities, and limited resource farmers?

Extension is finding itself pulled in two directions—to reach out to persons with specialized needs while, at the same time, continuing to serve more traditional audiences. Some critics object to Extension concentrating too much on middle-income, rural, and farm residents, while others would like to see the clientele of Extension more narrowly defined in terms of the more traditional audience. However, before one gets caught up in arguments for or against a position of expansion or retrenchment, it is crucial that we have a factual basis for establishing just who Extension's clientele really are. There is an absence of a policy statement defining who is the appropriate target audience of Extension. Without this clearly stated goal, it is impossible to assess organizational performance.

"Extension Clientele Are Satisfied with What They Receive"

It has been said that Extension runs a quality program, that it is one of the few programs developed out of an expression of local needs, and that it is based on sound research findings. This must be true to some extent, or people would not continue to participate in a voluntary program. As Jenkins concludes, "When it is all said and done, after all, it is the student or client, and no one else, who takes or leaves what a voluntary educational agency has to offer, thereby rendering effective judgement on its programs" (1980:1). However, evaluation efforts have done little more than count participation in past years. Other than in informal ways, clientele have not been asked how satisfied they are with the service they are receiving.

As with any public agency, dissatisfaction with Extension does exist. Individuals may have ceased going to Extension for assistance because the organization just does not serve their needs. Some would argue that Extension agents are generalists and are not able to keep up with the latest developments. While others would contend that Extension's agricultural information is only appropriate to large, commercial farming operations or that home economics information is geared more to promoting the concerns of retailers than the concerns of homemakers.

Inasmuch as we do not have a widely accepted mechanism for gauging the level of satisfaction with Extension programs, it is impossible to determine whether the degree of satisfaction increases or decreases as changes in programs are implemented. There are no standards for comparison.

"Farmers Represent Extension's Primary Base of Support"

It is generally concluded that Extension's primary support base comes from agriculture. From the start, Extension has looked to farmers and farm organizations to represent the interests of the organization. Smith and Wilson (1930:21–22) observe that "the task of securing . . . funds for cooperation with state and federal government is largely left to the farmers." However, we have also seen the influence homemakers can have on their elected representatives. They have often mobilized letter-writing campaigns and have successfully lobbied elected officials on behalf of the Extension budget. Others point to the 4-H program and claim it as Extension's real support base. After all, the development of youth can be a very emotional issue. And, then, what about the influence of community leaders who have been assisted through the community development program? Because Extension is locally based, it is easier for citizens to see the benefits of the program. For these people, the Extension organization is the county Extension staff.

We are in a period of disenchantment with government and its representatives, such that less government is seen as desirable. The best that most programs can expect is a continuation at existing levels.

The critical question for most administrators is how popular support translates into political action. Organizational goodwill may not directly translate into budgetary support. In times of limited resources, concern ultimately is on how to influence the political process in the allocation of dollars.

One would expect those persons who use the services of Extension to be more supportive of it. If a person receives benefits from a program and is satisfied with the services, then, this feeling should translate into a willingness to expend resources for its continuation. Whether, in fact, it does has not been confirmed.

"Extension Staff Rely on One-On-One Contact for Reaching Clients"

Seaman Knapp set the stage for the use of farmer demonstrations as a primary Extension method in his work with cotton farmers in Louisiana at the turn of the century. Extension has since continued to rely heavily upon individual and small group methods that emphasize decision making at the individual level. Demonstrations are still utilized, though home visits, telephone calls, and office visits are the predominant individual methods utilized today. The Extension agent also has available a large array of printed publications that serve to supplement personal communications. Extension staff make extensive use of such group methods as workshops, leader training meetings, and subject matter meetings. However, in dealing with concerns of a public nature, there has been increased focus on decisions at the community level. In recent times, Extension has made increased use of mass media communication methods such as radio, TV, and newspapers; and some would contend that Extension needs to explore new ways of using modern telecommunication delivery systems (Hildreth and Armbruster, 1981). This could include the use of computers, videotext, teleconferencing, and the like.

Most people would agree that Extension staff have been able to reach a larger number of persons through the increased use of mass media and group methods in place of one-on-one contacts. At the same time, there are programs that require the use of individual program delivery methods—for example, the Expanded Food and Nutrition Education

Program, small farmers, and urban youth. The educational methods used by Extension are generally thought to be effective; however, has the relative impact of the different methods ever really been examined? What is the best mix between individual, group, and mass media methods for attaining the greatest impact and programming efficiency? Under what conditions should each be used and for what types of clientele? Which do clientele prefer?

Issues Facing Extension

A number of policy issues arise from the issues presented in this chapter. Some individuals see Extension as moving too far away from its traditional programs and clientele. In this context, "traditional" generally refers specifically to production agriculture, homemaking skills, and similar farming and home economics subjects for youth. In fact, in some circles Extension is still trying to justify its home economics and 4-H programs. More "socially oriented" programs are seen as deviations from Extension's intended mission. Others conclude that Extension should, first and foremost, respond to local needs. And it is this expression that should guide Extension's future, not federal and state directives. Traditionalists imply that "program changes [that] have largely mirrored national trends or have reflected changes in the demographic, economic and social characteristics of the population" (U.S. GAO, 1981:10) are inappropriate reasons for altering program directions. In order to begin to resolve these differences in expectations, the purposes and intended audience for Extension programs need to be examined in the context of today's needs. Most everyone agrees that a clarification of Extension's mission statement is both necessary and desirable, but the problem results in trying to reach consensus on its content.

Improved evaluation and accountability in Extension are called for in the Congressionally mandated evaluation, the ECOP Task Force on Evaluation and Accountability, and the U.S. GAO report. Though there may be differences of opinion as to what types of evaluation are desired, there is no lack of enthusiasm for the concept. The real test will come when it is time to commit resources to such efforts.

The concept of organizational legitimacy is one that Extension has largely ignored. Extension's philosophy has been that a quality program will sell itself and that it is inappropriate to promote the organization. However, with recent threats to Extension's resource base has come a realization that if the organization is to survive it must address issues of political and social legitimacy. Administrators, as well as field staff, will need to devote more time and energy to understanding the process of establishing and maintaining legitimacy with the environment.

Notes

1. Farm Population: 1910–1920, *Historical Statistics of U.S. Colonial Times to 1970*, p. 457, Series K1-16; 1930–1975, *Statistical Abstract of the U.S. 1978*, p. 685, no. 1164; 1978, USDA, *Agriculture Handbook*, No. 561, p. 37. A farm is any place with annual sales of $1000 or more. Years 1960–1978 based on 1959 Census of Agriculture definition—a farm is a place of 10 acres or more with agricultural sales of at least $250. In the 1950 and 1954 Censuses, places of 3 or more acres were counted as farms if the annual sales amounted to $150 or more. For definition used in earlier censuses, see *U.S. Census of Agriculture: 1964*. Alaska and Hawaii were included from 1960 to 1978. Rural Population: Historical Statistics of U.S., Colonial Times to 1970, p. 11, Series A57-72. *Statistical Abstract of the U.S. 1978*, p. 17, No. 14 and *Data Users News*, Vol. 16, No. 9, September 1981. In Censuses prior to 1950, the rural population comprised all persons not living in incorporated places of 2,500 or more. In 1950 and 1970 the definition was modified somewhat.

2
Evaluation in Extension

Systematic evaluation is increasingly sought to guide operations, to assure legislators and planners that they are proceeding on sound lines, and to make services responsive to their publics.
—Cronback and Associates, 1980:12–13

Evaluations are expected to answer many questions. They are utilized by a variety of individuals for a variety of purposes. Program managers are interested in evaluation for purposes of program intervention. They are concerned with the relative effectiveness of different strategies or methods and their expected outcomes. Such information can then be used to modify the allocation of facilities and resources among existing programs for purposes of increasing effectiveness. Organizational managers focus on structural arrangements that are under their control. These include such aspects as rules and regulations, standards of production, division of responsibility, and supervision.

Employees emphasize process measures of evaluation because they do not fully control production outcomes. Since they have limited influence over the selection of work activities and outcomes, they tend to focus on "how" they perform in relation to specified standards. This is especially true for Extension in that studies of Extension are generally initiated by staff members for purposes of improving program performance.

For others, evaluation is a tool for cost analysis. Efficiency is seen as an internal standard of performance that relates the output produced to the resources used. This method provides an indicator of program costs relative to alternative uses of resources. Efficiency measures are gaining in popularity as decision makers face the dilemma of allocating scarce resources among many alternative uses. There is a desire to make optimum use of the resources available, to produce the most bang for the buck.

Effectiveness is used as an external standard of how well the organization is satisfying the demands of those outside of the organization.

The environment of the organization is composed of a variety of individuals and groups, each with its unique set of demands and expectations that are communicated to the organization through feedback mechanisms. Effectiveness is measured in terms of the attainment of these environmentally defined outcomes. In other words, do clientele feel that their needs have been met? In addition, clients place value on how well services are delivered (i.e., were staff prompt and courteous?). Representatives of the public-at-large, on the other hand, are likely to be more interested in what Scott (1977) refers to as the indicators of "macroquality." Is the organization concentrating its attention and resources on the proper products or problems? Is the community as a whole benefiting from its operation?

An evaluation of organizational effectiveness is the judgment of certain individuals or groups. These evaluations may include persons internal to the organization as well as from the organization's environment. Each person holds his or her unique perspective that may differ, or even conflict, with that of another. Organizations seldom satisfy all client groups, let alone all interested parties. Consequently, compromises are made and certain viewpoints are given precedence over others. Generally, organizations like Extension that face more needs than can be adequately served tend to satisfy various constituencies to some degree, though not satisfying any completely (Cameron, 1981). Some services are allocated to most all groups. Under these conditions, effectiveness needs to be measured from multiple domains. What may appear to be ineffectiveness from a single point of view may be very effective when viewed from a total perspective.

Organizational Effectiveness Examined

The many views of organizational effectiveness can generally be classified as either a systems or a goal approach. The systems perspective emphasizes the relationship of the organization to its environment and the ability of the organization to acquire and to maintain adequate resources that enable it to function (Yuchtman and Seashore, 1967). In the systems approach, the organization is viewed in an exchange relationship with its environment. Actors within the environment either directly or indirectly supply the resources for organizational inputs and are the recipients of the outputs (Pennings and Goodman, 1977). The environment reacts to the performance of an organization like Extension by controlling the allocation of resources; it may choose to continue to support the organization at existing levels, to expand the organization by increasing support, or to decrease or to discontinue its support. The environment is a political arena in which constituents sanction the

FIGURE 2.1
Open Systems Model

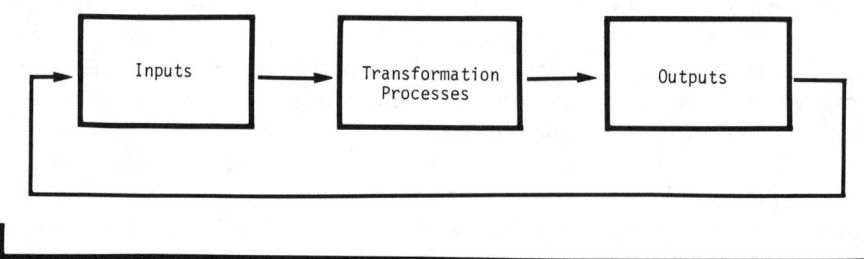

External Environment

allocation of resources, influence the establishment of constraints, and define the criteria for determining effectiveness. Thus, to a great extent, constituents decide the criteria used for assessing effectiveness and make judgments as to whether Extension performs adequately.

The concept of the systems model includes three elements: inputs, program operations, and outputs (Figure 2.1). Inputs represent the factors that are "invested" in the organization. These inputs are then transformed into outputs that are returned to the larger environment (Steers, 1977). The systems model defines the pattern of interaction of these elements.

The goal approach, on the other hand, focuses upon the performance of the organization in the attainment of certain specified ends (Etzioni, 1964; Price, 1972). These goal statements provide the criteria for determining effectiveness in that the more closely output approximates the organization's goals, the more effective it is thought to be. Goal statements are generally taken from formal mission statements of the organization; however, Etzioni (1964) contends that the "real" goals should be determined by asking the organizational participants. While the goal approach appears very straightforward, deciding on the "desired state of affairs" has created problems in determining effectiveness. Extension has a multiplicity of goals that are sometimes conflicting, and often nonquantifiable, abstract statements of general purpose that vary according to who defines them.

Most goal-oriented models assume that organizations eventually reach consensus as to the goal or end to be pursued when that is seldom the case. In reality, different expectations exist; one individual may see the organization as effective, while another views it as ineffective (Scott et al., 1978). From whose perspective should the assessment be made? Is it the perspective of the employees, administrators, clientele, or the public-at-large that determines the appropriate criteria? Among private firms, little agreement was found among organizational members and customers as to their assessment of effectiveness (Friedlander and Pickle, 1968). And, yet, the goal approach has generally relied upon statements specified by organizational managers. Utilizing environmentally defined goals could generate quite different findings.

No single approach for evaluating organizational effectiveness is appropriate in all circumstances. Organizations can be judged effective on certain criteria and ineffective on others, or the judgment may differ on a single criterion by different evaluators. For these reasons, it is important to utilize a variety of approaches that draw upon different types and sources of information.

Supporters of both the systems and goal approaches agree that each position has limitations. The systems approach is seen as too abstract and difficult to put into operation, whereas the goal approach faces problems of specification and the resolution of differences of desired consequences (Kahn, 1977). Because neither seems adequate in itself, a compromise is in order. Cummings (1977:61) argues that, "Organizations are best assessed as instruments of outcomes; that is, the effective organization is the organization that best serves those who perceive it as a means to their ends." Thus, organizational effectiveness is defined on the basis of intended outcomes (goals) by constituents within the environment (system). This approach recognizes the importance of the environment in providing organizational legitimacy and, at the same time, allows recipients of the service to assess performance based upon their own expectations for the organization (Christenson and Warner, 1982).

In educational organizations like Extension in which outputs are difficult to quantify, effectiveness is often represented by their institutional image. Goal achievement may be seen as a desired end, but the vitality and support for a public agency may be as much a function of political agreement and the social definition of the organization as the outcomes achieved.

The Input-Output Approach

Miller's (1979) modification of the traditional input-output approach contains the same elements as the systems model. It relates input to program operations to outputs. None of these three elements is seen as sufficient in themselves but must be examined in relationship to the others (Figure 2.2).

FIGURE 2.2
Three-Stage Model of Program Evaluation

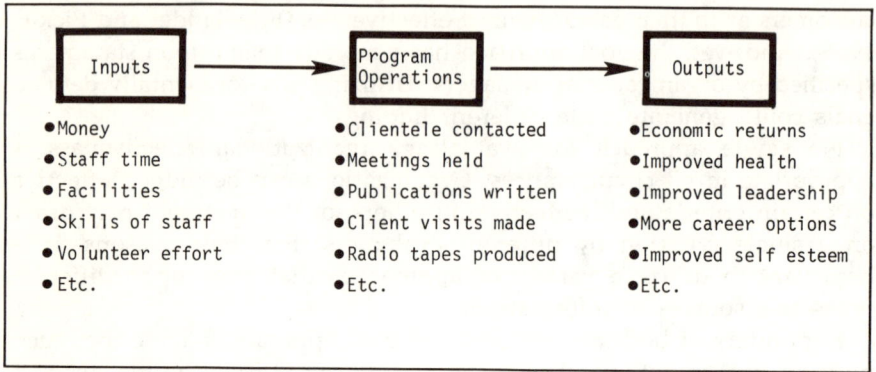

Adapted from Miller (1979:65).

Inputs

Inputs are what make a program possible. They are the resources that go into a program's development. Inputs include such things as staff time, skills, funds, and facilities. The adequacy of inputs generally comes into question when appropriation bills are before the legislature, when budgets are prepared, and when programs are funded. However, determining when resources are adequate is a difficult task in that program administrators often indicate they are inadequate, no matter what the level of support. Adequacy of inputs can only be assessed in relation to the level of program operations.

Program Operations

Program operations are what some have referred to as program activities or efforts. Evaluations of operations focus on "the quantity and quality of activity that takes place" (Suchman, 1967:61). They focus on what is done, and how it is done. Was a program carried out? As Tripodi et al. (1971:45) observe, "Evaluations of program effort refer to an assessment of the amounts and kinds of program activities considered necessary for the accomplishment of program goals." The measurement of operations is generally accomplished through an organizational monitoring or reporting system.

Extension has spent a considerable amount of time and effort on its monitoring system—the Extension Management Information System. EMIS supplies performance level measures for determining the extent to which target populations are being served, the quantity of staff performance, and compliance with established standards. Probably the most frequently used indicator of the level of program operations from EMIS is the number and type of participants served by the program (clientele). However, use of the monitoring system needs to be broadened to include such things as organizational processes and educational methods.

Program activities are a necessary but not a sufficient condition to insure outputs. In other words, there cannot be outputs without program operations, but operating a program does not guarantee there will be results such as changes in clientele behavior. Relating input to output is a common practice in evaluation modeling, but Miller (1979) argues that what is often missing is the consideration of program strategies for converting inputs into outputs. It is what Weiss (1972:43) calls the "black box" approach to evaluation. How one converts inputs into outcomes is left a mystery. Without the specification of these activities, there cannot be an adequate understanding of why programs succeed or fail.

Outputs

Evaluation research in the field of formal education is dominated by an emphasis on measuring outcomes (Provus, 1971). Actual outcomes achieved are compared with desired outcomes (goals). In education, these outcomes are measured in terms of the performance on achievement tests; in law enforcement, crime rates are used as indicators of outcomes; and, in health programs, outcomes are often seen as changes in the incidence of diseases. These are seen as the results of the efforts or, to what some refer, as the impacts, effects, or consequences. In Extension, the type of evaluation that is most lacking is the measurement of program impact. Being an informal educational program, documenting program impact has been difficult. As a result, the monitoring of program operations has been used as if it were an indicator of output. That is especially evident in the 1977 Congressionally mandated evaluation. The purpose of the evaluation emphasized the "consequences" of Extension programs, but the bulk of what is reported stresses the level of participation. As Suchman (1967:60) concludes, "Whether a public service program can be established (effort) is quite a different question from whether it does any good (effect)."

The task of measuring output from educational agencies is much more difficult than measuring output of private firms. Quantities produced and/or manufactured can physically convey outputs from private enterprise (e.g., number of clocks produced). However, in educational institutions, the products are not as clearly defined. Thus, in measuring effectiveness, one focuses on clientele to determine the quality of the product. Since clients purchase physical products in private enterprise, the number of units sold provides a quantitative assessment of client reaction. Holzer (1976) argues that, because of the unique nature of public organizations, effectiveness in such organizations should be measured in terms of client satisfaction with the organization's goods and services. He points out that such perceptive or qualitative indicators serve a similar role as the sales and profit information for the private sector. Thus, in many ways, whether one is dealing with a public agency or with a private industry, client reaction is a crucial element in evaluating agency outputs.

Evaluation studies of public agencies have tended to neglect outcome measures (Scott, 1977). It is logical that the emphasis would be on topics of interest to organization managers and workers in that they are the ones who commonly identify the need for evaluation and the resulting topics to be examined. Indeed, they are the persons deciding upon the funding of such efforts. Also, most of the data available for analysis is collected by the organization itself. This information generally focuses upon operational aspects of the organization; whereas outcome data have to be collected by the evaluator with a substantial investment in time and money.

Systems Effectiveness Model for Extension

The Systems Effectiveness Model as presented in Figure 2.3 is our approach. It recognizes the concepts of inputs, program operations, and outputs of input-output models but also integrates organizational and environmental dimensions of the systems approach. The general conceptual framework was developed by Etzioni in 1964 as an alternative to organizational evaluation by the traditional goal approach. The systems effectiveness approach constitutes a statement of relationships among different components that is necessary for the organization to maintain itself and to operate effectively. It recognizes all of the aspects of the organization within its environment and suggests their interrelationship. The following discussion specifies each of these elements for the Extension organization.

Organization

Starting at the upper left hand corner of the diagram, the Extension Service (1) represents an organization grounded in agreements between federal, state, and local governments. The organizational purposes of Extension are stated in the Smith-Lever Act and its amendments over the years. As the educational arm of the land-grant system, Extension is charged with the responsibility of "extending" knowledge beyond the walls of the universities. Though the state and federal partners have influence over the direction of the organization, Extension operates in

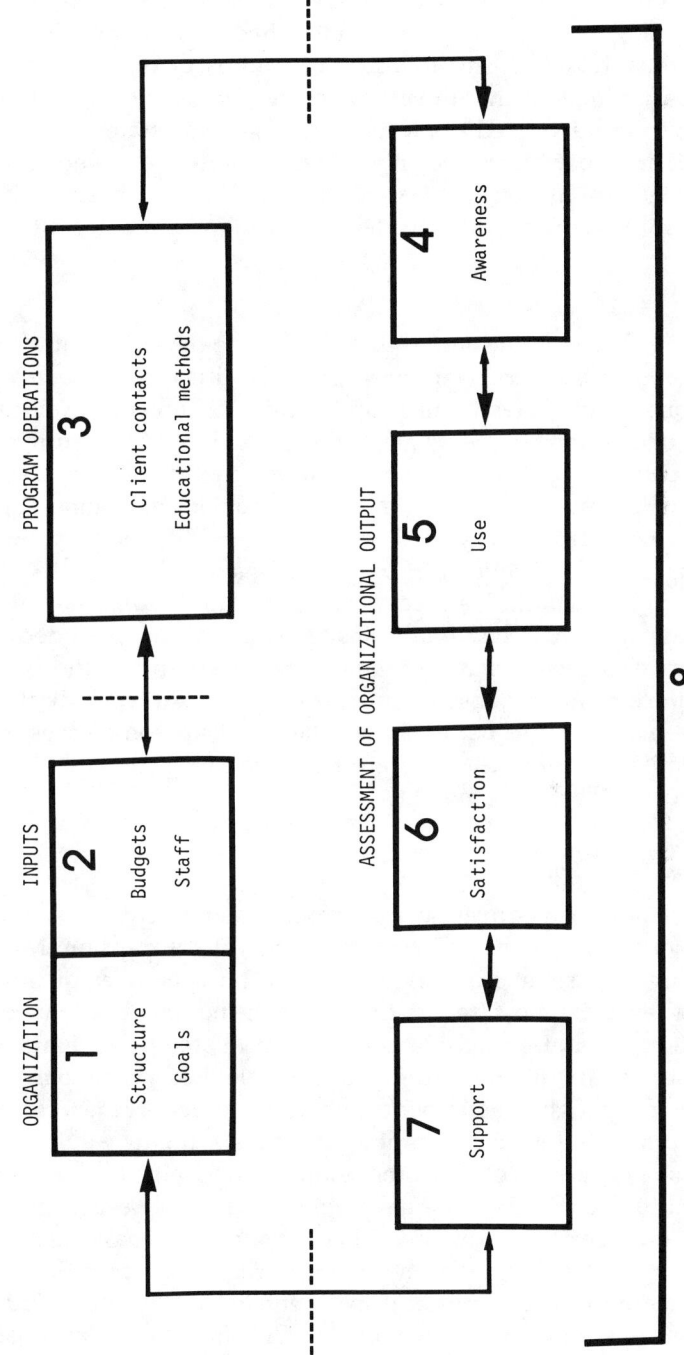

FIGURE 2.3
Systems Effectiveness Model for the Cooperative Extension Service

each county of the United States by providing programs, information, and training to meet the needs as identified by local people. Extension's goals, broadly stated, are to provide services in the application of the research findings of the university to the problems of agriculture, home economics, and related subjects. These general statements have been operationalized in terms of more specific objective statements on such diverse topics as agricultural marketing, human health, community services, livestock production, leadership development, and others.

Inputs

Inputs (2) are the resources that go into the development of programs. They are represented by money, staff, facilities, and the like and are appropriated by federal, state, and local governments. Necessarily, assessments of inputs are generally restricted to those things that are easily measured. This does not allow for the qualitative assessment—only quantitative. For example, it can provide an enumeration of the total number of staff assigned to a program, but it says nothing of the training and qualifications of these persons. Or, in the case of physical facilities, the quantitative assessment may indicate that they are present; whereas they may be totally inadequate for the intended use. The qualitative elements may be equally as important as the quantitative. Nevertheless, in the absence of adequate qualitative indicators, quantitative measures are better than none at all. In some cases, those that are relatively easily measured are not even being utilized. At least they can provide a starting point.

Program Operations

Inputs are then converted into programs and activities. The program operations (3) are the efforts generated. The Extension Management Information System provides useful data on a number of indicators of activity level; the most frequently used is the number of clientele contacts. This figure includes multiple contacts with some individuals; therefore, it is not the number of different clientele who have participated in programs. Though the number of contacts does reflect the total magnitude of the program, the actual number of different clients would be expected to be somewhat smaller. The other indicators examined are the educational methods utilized in the program delivery. We will be especially interested in the differential use of individual, group, and mass media methods as they are related, on the one hand, to the inputs of expenditures and staff time and to the appraisal by clientele, on the other. Indicators of program operations in Extension originate from organizational records.

Awareness

Though not all of the public use the programs of Extension, a large percentage would be expected to be aware of their existence. Most evaluation models limit the study of clientele to consumers of the services (Katz et al., 1977); however, the viability of an organization does not rest with consumers alone, even in the private sector. Because public agencies depend upon the legislative process for funding support, the survival of the organization can be determined as much by those who are unaware of the organization as those who are aware. Therefore, though it may be inappropriate for persons to rate the services of an organization they do not use, it is important to assess their general level of awareness of the organization. This level of awareness (4) is determined by asking people whether they have ever heard of the Cooperative Extension Service or its programs.

Users and Nonusers

Programs of Extension are directed at the general public, but, because of Extension's diversity, the specific target audience varies by program. However, it has been said that there is something for everyone in Extension. Nevertheless, some people choose to use the services of Extension, and some do not (5). So, we end up with a certain proportion of the population that consider themselves clientele of Extension.

Identifying the clientele of the Cooperative Extension Service would appear to be a relatively simple task; however, in reality, it is not so straightforward. Usually, the clients of an agency are defined by enrollment records, mailing lists, and eligible participants, but such methods do not provide a complete listing for Extension. Many other persons who never appear on the agency's lists are reached through the mass media methods of radio, television, newspapers, and magazines. The only method for identifying those individuals is to ask them. This approach, however, requires some awareness of the organizational identity to enable the individual to associate the service with the organizational name. This is made especially difficult with some of the mass media presentations that do not identify the material as originating from Extension. Therefore, one could expect any such indicator to provide a conservative estimate or, in other words, an underreporting of reality.

It is also difficult to control for the time dimension of use. Was use within the past year, the past five years, or twenty years ago? And, then, how accurate is the recall over an extended period of time? In addition, there are problems in getting an accurate assessment of use by nonadults, especially since the 4-H program is directed primarily at children ages nine to nineteen. Nevertheless, carefully designed questions asked of the right individuals can serve to identify the clientele of Extension based on the citizen's own perception of use.

The intended clientele of Extension could be broadly defined as anyone not a student at a land-grant university. A more pragmatic approach, and one to which Extension has generally subscribed, is to define the appropriate audience based on those individuals who have need for the service. In order to identify those persons, this approach would seem to require a fairly comprehensive needs assessment. However, as an easy way out, it has merely been assumed that certain groups of individuals such as farmers, youth, and homemakers have a need for the service. For example, for the agricultural program to define its client group as agricultural producers is both misleading and inadequate. Such an approach is too simplistic, because, first of all, not all farmers would want to be considered potential Extension users; and secondly, many present users (homeowners, agribusinesses, and gardeners) are not even included in this definition. Target audiences must be specified in more accurate and realistic terms, because, without clearly defined expectations, an assessment of effectiveness on the basis of utilization of the service is extremely difficult, if not impossible.

Satisfaction

To measure impact is "to establish, with as much certainty as possible, whether or not an intervention is producing its intended effects" (Rossi et al., 1979:161). The most widely accepted method is to determine the extent of goal attainment. However, that approach has inherent problems. Clearly defined measurable goals are generally lacking; and, even if they are available, it is difficult to determine the net effect in outcome that is attributable to the effort of the program. Because of difficulties in measurement, output measures are often restricted to indications of economic and physical impacts. An additional complicating factor with educational programs is that outcomes are often not immediate behavioral effects but, rather, long-term changes in attitudes, knowledge, and personal development. In the absence of easily quantifiable measures of output, there has been increased attention to qualitative assessments by the recipients of the service. Important sources of information that have frequently been overlooked are the perception, attitudes, and experiences of people in their association with public agencies (Christenson and Taylor, 1982). This places the attention on the appraisal of the service by the user. As Cummings (1977) points out, maybe the best indicator of organizational effectiveness is how well it serves those who have a need for its services. The differences between quantitative and qualitative measures are exemplified by contrasting the number of patients treated in a clinic to the feelings of satisfaction with the quality

Evaluation in Extension　　　　　　　　　　　　　　　　　　　　　　　　37

of the health care, or by contrasting the crime rate for a city compared with the residents' feelings of security.

Satisfaction (6) with public agencies is increasingly used as a qualitative measure of organizational effectiveness (Katz et al., 1977). This indicator provides an assessment of the organization through the eyes of the consumers. Though qualitative indicators such as client satisfaction do solve some of the problems of quantitative measures, they, too, have limitations. An indicator of general satisfaction with a program or agency does not specify the sources of the satisfaction or dissatisfaction. More detailed followup questions are necessary to secure that information. A second problem is in the determination of acceptable or appropriate levels of performance. Traditionally, comparisons have been made between levels of satisfaction reported by different social and economic groups. More recently, Katz et al. (1977) suggest expected levels of satisfaction for different government agencies. Looking at a wide variety of government service agencies, they find that approximately two-thirds of users are satisfied with their experiences with government agencies. However, they do observe that such a percentage of satisfaction should not necessarily be accepted as the standard of excellence; some agencies may require a more ambitious standard, while for others it may need to be lower. Nevertheless, their findings provide a basis for comparison with similar public service agencies and establish a baseline for assessing change over time. In addition to the information on the overall level of satisfaction with the Extension organization, it is also possible to obtain similar data for the four main program areas.

Support for Extension

Public. The next component of the model is a consideration of the relative funding support desired for Extension (7). Is the public generally positive about the tax dollars being spent on Extension programs?

In examining how satisfaction with a service translates into public support for the agency, Katz and associates (1977) find that a person's specific experience with a public service is more positive than his or her general attitudes about the agency that delivered it. The general feeling toward the agency is negative if a person has a specific experience with the agency that is negative, but a positive experience does not necessarily translate into a positive attitude toward the agency. In other words, "A negative experience with an agency lowers one's general evaluation of government, but a positive experience does not raise it" (Katz et al., 1977:186). Persons with positive experiences and persons with no experiences with an agency are found to give essentially the same expression of general support, while those with negative experiences give lower general readings. Therefore, specific experiences are more likely to transfer into a person's general assessment when they are negative.

Organized. Probably no other government agency has as extensive an involvement of local citizens in the program development process as does Extension. Advisory councils or committees are frequently composed of people who are involved in Extension programs and/or are influential community leaders. Though the expectations of these groups vary by county and state, they generally have input in the determination of program directions, the selection of staff members, and the preparation of budgets. And, in times of limited resources, these advisory committee members have also served as spokespersons for the Extension organization in governmental appropriation processes.

In addition to committees constituted specifically for the purpose of advising and supporting the Extension program, a number of other groups are closely related to Extension and are affected by its vitality. In agriculture, there exist farm organizations such as the Farm Bureau Federation, the Grange, the National Farmers' Organization, the National Farmers' Union, and the American Farmers' Movement. There are also many agricultural commodity groups such as the Pork Producers, the Cattlemen's Association, the Soybean Producers, the Milk Producers, etc. These farmer organizations not only have influenced the nature of Extension programming, but they have also provided an effective lobby in support of Extension. As Alfred True (1928:165) wrote over fifty years ago, "The college has no technical or official connection with the state federation of the farm bureaus but hopes for its support in carrying out the college program in the maintenance of soil fertility and matters resulting from research and in securing funds for the educational and research work of the college."

The agricultural community, as broadly defined, has been seen as the primary base of organized support for Extension. Though most would agree that Extension's base of support has broadened as its clientele has diversified, nevertheless, many Extension administrators still point to the fact that, in times when the organization is under fire, it still falls primarily to these agricultural groups to rally its support.

In addition to agricultural groups, other organized Extension clientele have been supportive of the organization. These include the efforts of the Homemakers Association, the 4-H program, and community development groups. In that homemaker clubs report over 600,000 members, the 4-H program claims five and a half million youth and adult participants, and community development involves many community leaders, their impact should not be overlooked. In fact, some would argue that there is more visibility and goodwill—and ultimately support—from these programs than there is from agricultural programs.

Funding

Funding for Extension is based upon a climate of political legitimacy and organizational effectiveness. Political legitimacy reflects an expression of relevant support groups, sentiments of the public, documented contacts, fiscal constraints, formula allocations, and leader endorsements. In this context, the image of Extension is often more important than evidence derived from evaluation studies. Governmental funding sources ultimately determine the resources available to Extension to be used as inputs. While it is unlikely that the Smith-Lever Act will be changed substantially to alter the basic structure, purpose, and scope of Extension, the allocation of financial resources is a very powerful tool for implementing change. Changes in the magnitude and type of funding have immediate impacts. With stable or declining resources, Extension, like other agencies, will have to improve its management or reduce the scope of its programs. Congress could also choose to change the balance of "formula" versus "earmarked" funds or alter the formula for distribution among the states. Any of these changes could have substantial impact on a state Extension Service.

The three-way partnership between federal, state, and local governments has proved to be an innovative idea. It is ironic that the concept of the cooperative funding arrangement as it was established at the turn of the century for Extension is now being utilized extensively for other programs. We are moving away from an expectation that the federal government can do it all to one of shared responsibility in program direction and support.

Environment

The interface of the organization and clientele must be related to the environment within which the organization operates, judgments are made, and legitimacy is established (8). Increased attention is being given the relationship of elements of the environment and organizational performance. Moos's (1974) social-ecological approach recognizes the importance of the context of program implementation in the determination of success or failure. The environment can be seen as including virtually everything outside the organization (e.g., technology, products, consumers, competitors, geographical setting, economic and political conditions), and they impact all parts of the model. In studying people's perceptions of the quality of community services, Christenson and Taylor (1982) conclude that people's perceptions are associated more with the situation in which the service is provided than with either the level of expenditures on inputs or performance measures of staff. If this is true, then much more attention needs to be given to the identification of relevant environmental factors and their relationship to such things as use, satisfaction, and support.

Sources of Data

Information for illustrating the model comes from several sources, the most important of which is a 1982 national telephone survey of the adult population. The survey was based on a random digit sampling frame containing all valid combinations of area code and control office code numbers for the contiguous United States. Usable questionnaires were obtained from 1,048 adults with an overall response rate of 70 percent. Nationally, 96 percent of households have access to telephone service and by using the random digit dialing technique, residents who have recently moved or who have unlisted numbers have an equal chance of being selected. This sample provides estimates of approximately plus or minus 4 percent for the overall contiguous United States population within a 95 percent confidence level. A more detailed description of sampling tolerance, data collection procedures, and a comparison of sample characteristics with the total population is provided in Appendix A. The survey provides information on awareness, use, satisfaction, and support for Extension.

A similar state survey in Kentucky, with sufficient sample size to provide county-level data (N = 11,015; response rate of 69 percent), also contains responses on use, satisfaction, and support for each of its 120 counties (Christenson and Warner, 1982). Indicators of level of program activities, staff time, county expenditures, and use of different educational methods are taken from the Kentucky Extension Management Information System and other agency records. In addition, information from several other state and county studies are included as supporting evidence.

Limitations of the Study

The strengths of the study also dictate its limitations. This study focuses on the Extension organization as a whole, not on specific components. As a result, it does not attempt to examine the unique features of each particular program or activity. To do so requires more in-depth information of a very specific nature. An overall study can suggest areas for more detailed investigations, but cannot supply all of the answers to small units of analysis.

This effort is national in scope in that it draws upon a sample of the total U.S. adult population. Consequently, conclusions cannot be drawn for small geographical areas. The national sample is not large enough to be able to generalize for states and counties. The Kentucky study is illustrative of a methodology that can provide state and county information.

This study is cross-sectional in nature, thus one cannot suggest what changes have occurred over time. That would require a longitudinal design. But the study does provide insight into the present situation and contributes baseline data for future comparisons.

Survey sampling utilizes aggregated individual responses that take on special meaning because they can be used to represent a larger population. The method is efficient, but one that must gain acceptance within institutional and political circles. With the emotional appeal of individual case examples and testimonials, by comparison, aggregated data seem cold and impersonal. It is easy to forget that those aggregate numbers are really many, many individual responses. It is important to know that farmer Jones feels strongly about an issue, but it is even more helpful to understand that the majority of farmers across the nation share the opinion of farmer Jones.

The survey approach to evaluation provides behavioral and perceptual information on a number of aspects of effectiveness. However, such methods do not attempt to measure indicators of economic and social change resulting from Extension programs. If the concepts of organizational image and legitimacy are important to Extension's future, then we must learn to use measures of perception and behavior along with such indicators as dollars saved and production increases. Multiple measures are essential for effective evaluations.

A Plan for Evaluation

With the increased attention given to evaluation, it is essential that we develop useful evaluation efforts that enhance the decision-making process. Not all evaluation studies have been designed with utilization in mind, nor have policymakers always been receptive to their use. As Wholey et al. (1976:46) conclude, "The recent literature is unanimous in annoucing the general failure of evaluation to affect decision making in a significant way." Conducting evaluation studies is one thing; getting the results integrated into decision making is quite another. Therefore, the first step in accomplishing this end is to make sure that evaluation efforts address issues of importance to the program manager or policymaker.

Evaluation is often carried out at one of two extremes—either as an academic exercise that is too general for direct application or of a specific program that is too limited for purposes of generalization. There is little or no thought given to how each fits into an overall scheme. What is needed is for the bits and pieces of information to be brought together in a model or outline of the expected relationships. This process approximates what is generally referred to as theory construction, though it is probably unwise even to use that word in this context because of the level of abstraction that the term usually represents for many practitioners. Actually, all that is suggested is that the various components of evaluation need to be examined in a total context of the organization or program, not in isolation. This, then, recognizes the reality of the existence of the various parts. The evaluation task is to fit them together into a model that begins to specify the necessary elements and their relationships.

After reviewing a number of evaluation frameworks that have been utilized in other circumstances, it is possible to identify what appear to be the critical elements for developing an effectiveness model for the Cooperative Extension Service. It was agreed from the start that the model would have to be comprehensive and be capable of treating the Extension organization as a whole. Too often evaluation studies in Extension have dealt with a single program or project. The second expectation was that the assessment model be capable of including multiple indicators of effectiveness. Individual measures when examined alone were not seen as providing an adequate assessment of the many possible aspects of effectiveness. And it was concluded that the evaluation model must recognize the importance of the organizational environment. In the Systems Effectiveness Model, environmental actors provide the appraisal of the adequacy of organizational output.

This chapter has presented a comprehensive model for evaluating the Cooperative Extension Service. It has identified eight components and specified their relationships to each other. The task in succeeding chapters is to illustrate the model with information from national, state, and county sources to determine whether the suggested relationships in fact do occur in order to provide answers to the questions posed in Chapter 1.

3
Public Awareness of Extension

Knowledge of the existence of government agencies is an important determinant of utilization
—Katz et al., 1977:183

You're damned if you do and damned if you don't! Public agencies face a dilemma. On the one hand, they have to make their services known. On the other hand, citizens get upset at seeing their tax dollars used on self-serving agency advertising. Administrators who are concerned about a public image and spend money to enhance it are open to criticism. However, administrators who do not create an awareness of their agencies may end up facing drastic budget cuts. This is one of the dilemmas faced by Extension.

All organizations and agencies whether private or public have a public image. In the private sector firms engage in elaborate and expensive advertising and public relations campaigns in order to establish a certain public image. Though Extension may not advertise per se, Extension does have an image. Its image has evolved over time through contact and familiarity with the agency and/or its programs. Extension's vitality in the future will rest with its ability to develop, maintain, and enhance a positive and viable public image.

Awareness precedes use of an agency's services and support for its existence. Awareness is seen as the top or wide mouth of a funnel; use is the next lower ring in the funnel followed by satisfaction and support. Thus, one can understand why the private sector allocates large sums of money to the creation of a positive image or awareness of their products. Managers of public agencies also need to be concerned with a public image, since awareness of the agency is the first step toward use and support.

The Consumer and the Cost-Bearer

Organizations have many publics: the consumer, the client, the nonuser, and the cost-bearer. Individuals served by an organization are its most obvious public. They are the clientele or the consumers of the service. However, there are also publics that are not directly served by the organization. These include persons who are aware of the organization and its programs but do not use them as well as persons who may be totally unaware of the organization.

Public agencies relate to taxpayers in that all citizens are cost-bearers, whether or not they use the services. Public agencies often provide services such as food stamps, which are paid for by the taxpayers, even though most of the population is not eligible to benefit from the program. Public organizations are supported by the citizenry and have a public image among most segments of the population. Many persons who have only a vague awareness of the existence of a public agency have formed an image of its identity. This image, however informed or uninformed, then influences the person's actions as they relate to the organization.

The public's image of an agency influences not only those who participate in its programs but also the allocation of public resources. Though some persons may not be clientele of the organization, they can and do influence the process of allocation of tax dollars. If nothing else, by not specifying other uses for the resources, they allow the organization to exist. Ignoring this large nonuser group may become especially critical in times of limited resources, inasmuch as in most cases they substantially outnumber users. In addition, those persons making policy decisions are often found in this nonuser category.

Public Opinion and Organizational Image

Knowledge about Extension occurs as people "experience" the organization as direct users of the services or indirectly through others. The image people have of the organization is developed through the direct and indirect experiences they have had with the organization and/or its programs. When people talk about Extension, they express both information about the organization as well as impressions and interpretations gathered over the years. Moreover, images are not grounded in fixed events but rather in information and interpretative processes that are constantly changing.

Knowledge is a set of symbols that serves two functions. First, knowledge provides a "relation to reality" by containing factual information (Etzioni, 1968:136). Second, knowledge gives meaning to actions. Meaning is derived from information we get from our environment through a mixing of factual information and interpretations. Individuals constantly update their knowledge through their own experiences and through the experiences of others.

Awareness as a component of knowledge is seen as including four elements: (1) complexity, (2) composition, (3) comprehension, and (4) comparison (Leff, 1978:114). The more complex an organization, the more difficult it is to develop a positive pattern of understanding. Thus, increased complexity may lead to unfavorable evaluations.

Composition is concerned with the assumptions we make about the organization because of its general nature. We have a preconceived meaning of what should be. In terms of public services, this involves understanding what a particular type of public agency represents, what services it is expected to provide, and how important it is to society. Such things as one's outlook on life and one's attitude toward public services in the community will color an evaluation of an organization.

Comprehension is related to the complexity and composition of an organization. For example, extremely complex organizations are likely to prove more difficult to understand, while a single-image organization is easier to comprehend. Some organizations, either in structure or name, provide clarity and ease of understanding. Others involve a multitude of identities. For example, Esmark Corporation is involved in a wide variety of different product lines and manufacturing processes. On the other hand, Gulf Oil is widely associated with oil-related products.

Finally, comparisons are important in that organizations are understood in a wider context. When we evaluate public services, we compare them with private services and other public services. When we think of Extension, we also think of other private and public service organizations that provide similar services.

Organizational research has generally focused on the internal workings of organizations. As a result, we know much more about such management concepts as span of control, delegation of authority, and specialization than we do about the organization's relationship with its environment. And what we do know about the relationship of the organization with its publics is usually limited to the users of the service or the buyers of a product. It was not until the 1950s that studies were conducted of the general public's opinion of government agencies (Janowitz et al., 1958).

Public opinion represents either a favorable or unfavorable expression of attitudes on subjects of public concern (Katz, 1960). Statements of public opinion provide broadly based expressions of the thinking of the population on a particular issue. In recent times, surveys have been used as a method for gauging the sentiment on selected issues. Such population surveys generally provide a statement of the direction and intensity of the opinion. The direction of opinion indicates whether a person is for or against an issue or object, while the intensity of the opinion registers how strongly a person feels. In addition, the existence of an opinion on an issue reflects at least a minimum level of knowledge about the nature of the issue; or, in other words, a common ground for discussion that undergirds and supports the differences of opinion (MacIver, 1955).

Public opinion should not be viewed as static and unchanging. Rather, every day brings new efforts to alter its nature. Methods used to influence opinion range from subtle, indirect techniques to sophisticated advertising campaigns. While most of the attention concerning the manipulation of public opinion has focused on marketing strategies in merchandising, the public sector also shapes its image in the eyes of the people. In fact, Edelman (1971) has suggested that the public sector has greater control over the perceptions of the public than does private interests, because government has fewer competitors in the forming of public opinion.

In recent years, we have heard extensive discussion of the silent majority, or what Dahl (1956) refers to as an indifferent majority and an intense minority. This situation not only reflects the relative inactive nature of the majority of the population but also explains the basis of special interest groups and their ability to influence public officials. We have tended to point to this "apathetic majority" as the reason why our educationally-oriented communications have not been successful. We tend to suggest that "somehow the targets of given messages are at fault for the absence of effect, rather than the creators of the content of the messages, or the media through which they are disseminated" (Mendelsohn, 1975:304). In short, communication research has shown that failures of communication systems generally can be attributed to the communicator rather than to the shortcomings of either the message or the audience.

The Image of Extension

As the size and complexity of society grows, the need for a public agency to communicate with its publics is more crucial than ever before. It is often assumed that a superior product or service will sell itself; that, if a person produces a better mousetrap, the world will beat a pathway to his/her door. Such is rarely the case. Even if a public agency delivers a high quality service, it is still necessary to communicate that message to the public.

In its early days, Extension programs were directed primarily toward agricultural producers, homemakers, and rural youth; therefore, public

Public Awareness of Extension

awareness of Extension was expected to be greater among those audiences. When most of the population were engaged in farming or lived in rural areas, Extension's thrusts were consistent with the needs of the majority of the population. However, with the decline in farm and rural population, the traditional audience of Extension is now a minority. With a broadening of Extension's constituency into urban areas, it is yet unclear whether Extension has established an identity among these residents as well.

With these new clientele has come a diversification of program thrusts. These changes are generally seen as positive responses to client needs, but, from the standpoint of organizational image, it is more difficult to project a single identity. These new programs have led to increased complexity and have made it more difficult for the public to comprehend the nature of Extension. Also of concern is whether the image being projected on behalf of the agency reflects this diversity, or whether Extension is still being described as exclusively agricultural in nature.

One would also expect the extent of awareness of a public agency to be related to the type of program it carries out and its methods of operation. Extension conducts an educational program that has relevancy to a broad spectrum of persons. It does not exclude persons from participation on the basis of eligibility criteria as do many social service organizations. Extension could be contrasted to agencies that serve a narrowly defined clientele (i.e., Veterans Administration). Moreover, Extension operates a program on the principle of extensive voluntary participation. Such a feature encourages widespread public awareness. Nevertheless, due to the unique subject matter of Extension's programs, it would be expected to be known to some persons more than others.

The purpose of this chapter is to address the issue of Extension's image with its public. First, it is necessary to determine the complexity and composition of its public image. Then, one can determine the degree of comprehension or awareness of this image. Third, one can establish some comparison level or standard for assessing whether the level of awareness is acceptable. Fourth, one can assess who is aware of Extension program areas. And, fifth, one can evaluate whether the organization is reaching its targeted and/or appropriate audiences.

Extent of Awareness

What percent of the U.S. population is aware of Extension? Estimates would likely vary substantially depending upon who is asked. Some might think it to be fairly low—about a quarter of the population because of the traditional rural focus of the organization. Others might say one-half based on the broad nature of its four program areas. Still

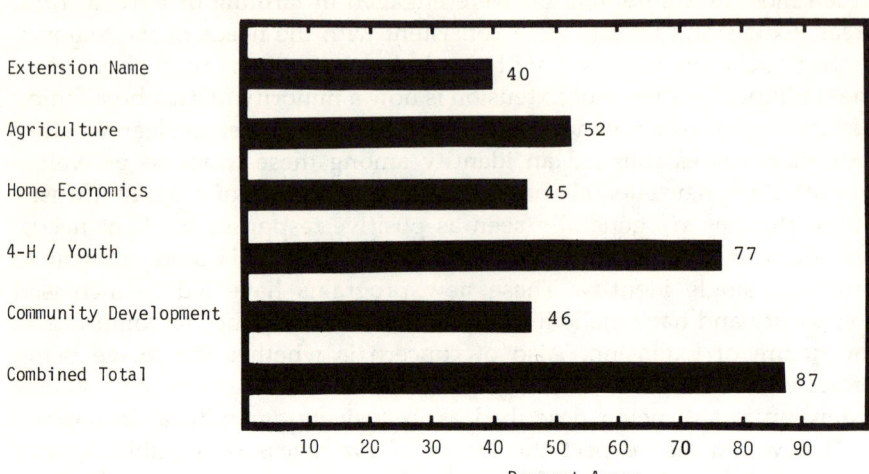

FIGURE 3.1
Awareness of Extension and Its Programs

others might optimistically estimate 75 percent. Our findings indicated that 87 percent of the population recognized Extension or its programs.

When a sample of the national population was asked, "Have you ever heard of the Cooperative Extension Service (sometimes called the Agricultural Extension Service) which is locally provided by County Extension Agents?" Forty percent said they had. In other words, slightly less than half recognized the name of the organization. Because of the diversity of state Extension organizations, the exact name may differ somewhat from the terms used in this question. It also could be argued that many clientele do not know the official name of the organization but identify it by such descriptors as the agricultural agent, the county Extension office, the 4-H agent, or merely by the names of the county staff. Others would point out that much of Extension's mass media information that is included in such outlets as farm magazines and television programs does not carry an organizational identification. All of these reasons would suggest that this figure represents a possible underreporting of the true level of awareness of Extension.

To test the notion that Extension is "known by many names," the aforementioned question was then followed by specific awareness questions on each of the four program areas—agriculture, home economics, 4-H, and community development (Figure 3.1). In all cases, more people recognized the program areas than the organizational name. A high of 77 percent had heard of the 4-H youth program, and about half recognized the other three. Because of its traditional mission, one might expect that

FIGURE 3.2
Percent Aware of Extension or Its Programs by Region

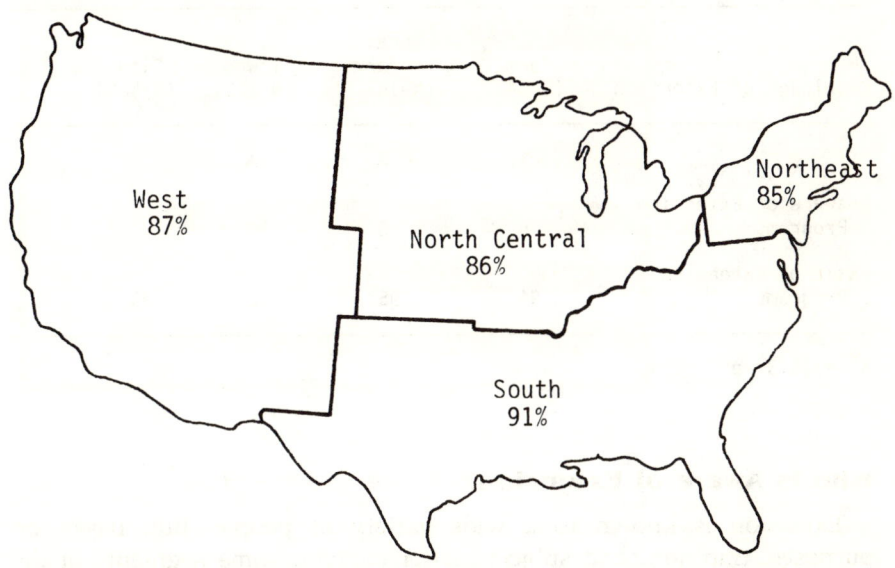

the greatest program recognition would be in agriculture and home economics. But such was not the case. The 4-H program had by far the greatest visibility. Of all of the four program areas, 4-H has the most widely identified name. It is short, easy to remember, and has not changed over time. The 4-H program also reaches a larger number of persons in diverse geographical areas. As might be expected, the second highest level of awareness was for the agricultural program with 52 percent. However, it is somewhat surprising that community development, the newest program thrust, had as much recognition as did home economics (46 and 45 percent, respectively).

A more accurate indicator of overall awareness would be to combine those recognizing any of the four different programs or the organizational name. Eighty-seven percent of the population indicated they had heard of one or more of the above. In other words, nine out of ten adults in the U.S. population were aware of Extension and/or its programs.

The level of awareness varies little in different geographical regions of the U.S. (Figure 3.2). It ranges from a low of 85 percent in the Northeast to 91 percent in the South. Given the diversity of programs in different states, there is a surprising similarity in the level of the awareness of Extension and its programs throughout the nation.

TABLE 3.1
Knowledge of Extension by Place of Residence

Knowledge of Extension	Farm (N=54)	Rural Nonfarm (N=163)	Town (N=292)	City (N=509)
	%	%	%	%
Unaware of Extension Programs	6	5	10	17
Aware of Extension Programs	94	95	90	83

$x^2 = 23.3$; $p \leq .05$.

Who Is Aware of Extension?

Extension is known to a wide variety of people. But, based on purposes, tradition, and subject matter content, some segments of the population would be expected to be more familiar with the organization than would others. The place people live is related to their level of awareness of Extension. As one would expect, persons from rural areas, farms, and small towns were more aware of Extension than were those who live in large cities (Table 3.1). Almost everyone who lived on a farm or in a rural area knew of Extension (95 percent). However, it is somewhat surprising that four out of five persons in the large cities also know about Extension.

When awareness was examined on the basis of other personal characteristics, significant differences were found with age, race, and the level of family income. As can be seen in Table 3.2, the lowest level of awareness was among persons with incomes of under $10,000, while the highest was for the income category of $30,000 and over. Awareness was about the same for the middle-income categories. Young adults (under 30 years old) were less aware of Extension than were all other age groups (Table 3.3). Black and other minority groups were the least aware of Extension when compared to other social, economic, or demographic subgroups (Table 3.4). Though differences were found among various segments of the population, one needs to keep these results in perspective. The lowest level of awareness of any of the subgroups considered was 79 percent; a level that would be envied by many organizations.

TABLE 3.2
Knowledge of Extension by Income

Knowledge of Extension	Less than 10,000 (N=262)	10,000-19,999 (N=262)	20,000-29,999 (N=251)	30,000- and over (N=246)
	%	%	%	%
Unaware of Extension Programs	17	10	15	8
Aware of Extension Programs	83	90	85	91

$x^2 = 9.26$; $p \leq .05$.

Awareness of the Four Program Areas

As was suggested earlier, awareness is related to the content of the organization's programs. However, not only does the level of awareness vary, but the characteristics of the persons who are acquainted with the programs also differ (see detailed table in Appendix B, Table B.1). Agricultural programs are least known by young people, persons with low incomes, and city dwellers and best known by persons of middle age with incomes of from $10,000 to $20,000 and with both high and low levels of education. Home economics is known to a similar audience in that young adults and urban residents are less aware of the program,

TABLE 3.3
Knowledge of Extension by Age

Knowledge of Extension	Less than 30 (N=340)	30-39 (N=226)	40-64 (N=335)	65 and over (N=116)
	%	%	%	%
Unaware of Extension Programs	17	11	10	10
Aware of Extension Programs	83	89	90	90

$x^2 = 11.18$; $p \leq .05$.

TABLE 3.4
Knowledge of Extension by Race

Knowledge of Extension	White (N=860)	Black (N=93)	Other (N=61)
	%	%	%
Unaware of Extension Programs	11	18	21
Aware of Extension Programs	89	82	79

$x^2 = 8.05$; $p \leq .05$.

while individuals with incomes of $10,000–$20,000 and over $30,000 and those with high and low levels of education are most knowledgeable.

Knowledge of the community development program takes on a different pattern than does agriculture and home economics. Those most knowledgeable of the community development program were older persons and blacks and other races. Contrary to the other program areas, there were no differences by place of residence, income level, and education. This means that the community development program is known equally well in urban areas as in rural and by all socioeconomic groups.

The 4-H program has the greatest level of recognition of the four program areas and also shows the largest differences in awareness by different population groups. Reporting substantially less knowledge of 4-H were blacks and other minority races, low income persons, and urban residents. Middle-aged persons and women were the most knowledgeable of the 4-H program.

When asked whether they had ever heard of the organizational name, the Cooperative Extension Service, 40 percent of the national sample said they had. This is less than half of the number that responded positively in the combined total as reported earlier (87 percent). Therefore, more than half of the persons did not recognize the organizational identity but were familiar with a specific program. Those who *did* recognize Extension by name were more likely to be in their middle years (40–64 years of age), white, in the higher income groups, college educated, and from a farm or the country.

Profile of Knowledgeables

Another way of presenting the information on awareness is to develop a profile of those persons who are knowledgeable (Appendix B, Table

B.2). Of those persons who know about Extension, half have a family income of between $10,000 and $30,000, with a small proportion having very low or very high incomes. This suggests that Extension is known primarily to persons with middle-level incomes. However, it should be pointed out that the income distribution of knowledgeables is almost identical to that of the general population (see Appendix A). Half of the knowledgeables have a high school education, while another third have some college (and 9 percent have advanced college degrees). Only 8 percent have a grade school education. Eighty-six percent are white, 9 percent black, and 5 percent are of other racial groups. Knowledgeables are distributed over the age groups the same as the total population. About half live in rural areas and towns, while half live in cities of over 50,000 people. Six percent live on farms.

Shoring Up Extension's Image

These findings reinforce the notion that Extension struggles with multiple identities. Multiple images provide a base that is not dependent upon a single client group; however, there is a lack of comprehension of the overall organization through which the programs are delivered. The potential for extensive organizationally based awareness is present. It is surprising that 87 percent, or nearly nine out of ten people, indicated that they had heard of one or more of the programs or the organizational name. It is up to the organizational managers to consolidate that identity. The private sector has done so through the use of a single corporate name or brand. It may not be necessary to go that far in the public sector, but the complexity can be reduced by avoiding the proliferation of different identities and by establishing closer ties between the different programs and the organization.

The private sector has successfully made use of public relations efforts in order to establish or modify its public image. Extension could learn from these examples. Because no organization, private or public, exists independently of its environment, it must continually strive to communicate its identity to the world around it. Existing clientele provide a valuable group of persons with whom to begin. In their association with others, these persons communicate an image of Extension—either positive or negative. And, if the organization wants to expand beyond these traditional constituents, it must establish means of reaching out to new target audiences. The basis of an expanded audience is present, since the findings demonstrate a high level of awareness among most segments of the population. There is at least a minimum level of awareness of Extension among almost all citizens.

To reach new audiences, the private sector has made extensive use of mass media advertising methods. Communication research indicates that mass media is limited as an instrument of persuasion (Klapper, 1960), while personal communication methods are seen as more effective (Katz and Lazarsfeld, 1955). However, despite the limitations of the media, when used together with interpersonal methods, they successfully transmit information to a large number of people in an inexpensive and rapid fashion.

Extension has a history of extensive use of personal communication methods; so much so that Extension has come under criticism for relying too heavily on one-on-one techniques (U.S. House of Representatives, 1981). It would be relatively easy, therefore, for Extension to increase the use of mass media methods in order to complement individual methods. In contrast, much of the private sector would find it very difficult and cost prohibitive to add this personal dimension.

Though the agriculture and home economics programs are fairly well known within certain segments of the population (especially rural residents), the 4-H program enjoys much wider name recognition throughout the total population. However, the 4-H program needs to be concerned with its lack of visibility among blacks and other minority races. In contrast, the community development program is equally well known by all types of people—urban and rural, rich and poor, black and white.

Awareness of Extension is an indicator of the image the organization is projecting to its publics. It provides a measure of its visibility and suggests the nature of future clientele. If a particular segment of the population is less aware of the organization and its programs, it is unlikely that they will become clientele or support the organization in the allocation of resources.

An important reason for examining the image of Extension held by the general population is to be able to compare their perception with that of organizational leaders and policymakers. Hasenfield and English (1974) point out that human service organizations tend to define their goals in ideological terms. There is agreement on the general principles, but differences often surface when these goals are translated into operational statements. Such appears to be the situation with the Cooperative Extension Service. Some federal Extension Service staff, congressional representatives, users advisory groups, and state Extension Directors feel that Extension's sole purpose is service to agricultural producers. However, when the general population was queried about Extension's image, they knew Extension primarily from their exposure to the 4-H program, not agriculture. This finding does not support the assumption that Extension is seen solely as an agricultural agency, at least not in the eyes of the general public. If the public's image of the organization differs substantially from that held by the agency, then either the organization is not successfully representing itself to the public, or organizational members are not reflecting the reality of the agency's programs. Either of which is reason for concern.

4
Who Are Extension's Clientele?

> *While the intended audience of public service agencies is generally defined as the citizenry, it is not at all clear whether the popular mandate intends service to 10 percent or 90 percent of the population.*
> —Katz et al., 1977

How many clients are enough? Should Extension try to serve every man, woman, and child in the U.S.? Who does Extension now serve? Are Extension clientele primarily farmers and rural residents? Are they rural youth and farm wives? Or are the majority community officials, homeowners, 4-Hers, and homemakers living in urban centers? Who should Extension be serving?

Extension finds itself pulled in two directions—on the one hand, to reach out to groups of individuals with specialized needs such as small farmers, minorities, the elderly, urban residents, and displaced homemakers while, at the same time, continuing to serve traditional farm and rural audiences. There exists a disparity between "traditionalists" who would like to see Extension pull back from the more socially oriented programs and services in urban areas and serve only farm and rural residents, and "expansionists" who desire that Extension make its educational programs available to all segments of the public regardless of their geographical, occupational, or socioeconomic background (U.S. GAO, 1981).

Examples of the stance of traditionalists can be found in the report of National Agricultural Research and Extension Users Advisory Board. In their report to the Secretary of Agriculture, they recommend that Extension redirect or eliminate programs and shift personnel to "serve the needs of producers of U.S. food and fiber" and that state Extension services "be directed to serve primarily the needs of the people of rural America . . . who do not enjoy the extensive social and public services that are available in cities and suburban communities" (1982:3 and 7).

In contrast, expansionists are critical of Extension for concentrating too much on middle-income rural and farm residents. They argue that Extension's target audiences have been too narrowly defined. Statements in support of an expanded role for Extension were given in testimony presented at Oversight Hearings on Extension conducted by the U.S. House of Representatives Subcommittee on Department Operations, Research, and Foreign Agriculture in March 1982. In these hearings, clientele were particularly supportive of the urban gardening program.

The Smith-Lever Act identifies the intended clientele of Extension as the "people not attending or resident in the land-grant colleges." Because the original purpose of Extension focused on the provision of educational information on agriculture and rural living, it is not surprising that the primary clientele have been farm and rural people. At least in part, the nature of the program content determines the clientele, and the clientele determine program content. With changes in the nature and size of the farm and rural population, in compliance with Congressional directives and in response to changes in clientele needs, the Extension program has broadened to serve a wide range, if not most segments, of the general public. For example, programs have been developed to address clientele such as urban residents with activity areas like 4-H, gardening, nutrition, recreation, energy, health, community services, and many aspects of family living. Specialized programs have been developed for the needs of pesticide applicators, and the Expanded Food and Nutrition Education Program was created to focus on the nutritional needs of low-income families. In short, Extension has become a multifaceted public service agency, which serves most segments of the general public on issues concerning agriculture, home economics, 4-H youth programs, and community development.

To a great extent, Extension has expanded into new subject and geographical areas without a comprehensive plan for the future. Rather, counties and states have gradually moved into new areas over time in response to local needs, availability of resources and personnel, and national priorities. While a few attempts have been made to identify emerging national issues and Extension's appropriate response ("The Kemper Report" in 1948, "The Scope Report" in 1958, "A People and a Spirit" in 1968, and "Extension in the '80s" in 1982), such efforts do not always substantially impact state programs.

Major changes have occurred in American society since the passage of the original enabling legislation for the Cooperative Extension Service. In response, Extension's programs have changed considerably. And yet, some individuals compare the Extension Service of today with the 1914 Smith-Lever Act and conclude that Extension has overstepped the intended mission. Such a conclusion is not so easily drawn. It is not merely a matter of turning back the calendar to the 1920s. Extension has responded to changing needs with new and different programs that service a different mix of clientele. Its success in doing so can only be assessed in the context of current needs.

Another client-related issue is the mandate to make services of public agencies available to the public without regard to race, color, national origin, sex, or age. Such statements evolve from Title VI and Title VII, the Civil Rights Act of 1964, and Title IX of the educational amendments and similar legislation. It should be noted that neither occupation nor geographical location are issues that have been singled out as criteria of discrimination in the provision of services. Thus, attempts to limit services to specific geographical areas do not appear to conflict with equal opportunity statements.

Extension's historical mandate has been to serve the public outside the land-grant colleges with special concern for agriculture and rural families. This statement has since been modified by Congressional acts, changes in population composition, and demands of specific segments of society. Evaluation of Extension's fulfillment of its overall mission and specific directives requires information on who is being served. Except for information collected for purposes of documenting equal opportunity compliance, Extension really does not know whom it is serving. Without this basic information, arguments that Extension's audience is either too broad or too narrow are meaningless. They are conclusions not based in fact but rather on speculation and commonly held assumptions.

Who Uses Extension?

What kinds of people use the services of Extension? Extension's primary method of documenting the nature of its clientele has been through an internal reporting system called the Extension Management Information System (EMIS). EMIS provides information to both state and federal governments on who is served by Extension. The information is collected by Extension staff and entered into a computerized data base. It reports the total number of contacts with clientele along with the race and sex of each. These contacts represent separate incidences of contact with clientele but not necessarily with different persons. Multiple contacts with the same individuals are included in these figures.

The EMIS system has provided information for documenting compliance with equal opportunity provisions and staff time distribution, but there has been only minimal use for management decisions. Part of the problem is that those close to EMIS reporting procedures question the quality of the data. Obviously, the quality of the conclusions drawn from the data can be no better than the information put into the system. If garbage is put in, you get garbage back out. Sophisticated analyses cannot improve the quality of the data. In addition, outside observers view information generated by agency personnel with suspicion. This type of information is often discounted on the basis of the potential biases inherent in an agency-based reporting system.

Studies that have focused on the identification of Extension's clientele are few in number. As part of the Congressionally mandated evaluation, the four program areas addressed some aspects of identifying clientele. Though somewhat limited in scope, the 4-H and home economics program areas conducted surveys of the general public (Gallup, 1979a; Gallup, 1979b; Pigg and Meyers, 1980). The community development evaluation focused on clientele identified by Extension staff (Mulford et al., 1980), and agriculture defined the target audiences as all agricultural producers. Earlier studies of users and potential users are also available, but they are usually limited to the farm population (e.g., Fuguitt, 1965; Nolan and Lasley, 1970).

To provide an independent assessment of the clientele of the Cooperative Extension Service, a national survey of the general population was conducted in 1982. The survey was not limited to a single program area; therefore, it provides a comprehensive picture of Extension clientele throughout the nation. The remainder of this chapter reports on the findings of the national survey and, thus, provides a factual basis for conclusions concerning the clientele of Extension.

Individual and Household Use

In the survey of the U.S. population, 23 percent of the persons questioned indicated that sometime during their life they had personally used Extension Service or contacted an Extension agent. In addition, approximately 20 percent indicated that other members of their family had used the service. The two questions are then integrated to give a composite measure of household use. In this combined total, 27 percent of all households in the U.S. had used Extension (Table 4.1). In most cases, if one member of the household had used the services of Extension, other family members had as well. This indication of household use may be a rather conservative estimate because the individual responding may have been unaware of the use of other members of the household. Another reason is that an individual may have received services but was unaware that they originated from Extension. Nonetheless, even if this is an underreporting, it remains that over one-quarter of U.S. households report using the Extension Service, nearly 22 million families. This national use pattern is similar to that of state studies in Kentucky,

TABLE 4.1
Individual and Household Use of Extension

Question: Have you personally _ever_ contacted an Extension agent or used the services of Extension? Have other members of your family _ever_ contacted an Extension agent or used services of Extension? (N= 1,028)

Response	Personal Use %	Family Members' Use %	Total Family Use %
No or Don't Know	77	80	73
Yes	23	20	27

Question: Within the _past year_ have you personally contacted an Extension agent or used services of Extension? Within the _past year_ have other members of your family contacted an Extension agent or used services of Extension? (N = 1,028)

Response	Personal Use %	Family Members' Use %	Total Family Use %
No or Don't Know[a]	90	91	86
Yes	10	9	14

[a]Those who said no to Question 1 were not asked Question 2 but were added (790 + 132) to the "no" category to get estimates of previous year's use.

25 percent (Warner and Christenson, 1981); Oklahoma, 37 percent (Cosner et al., 1980); Wisconsin, 27 percent (Wisconsin Extension Staff, 1979); and Missouri, 28 percent (Campbell et al., 1971).[1]

Because of the problems of recall and in order to secure an estimate of current yearly use of Extension Service, two additional questions were asked. First, we asked: "Within the past year have you personally contacted an Extension agent or used the services of Extension?" Approximately 10 percent reported personally using Extension during the

FIGURE 4.1
Household Use of Extension by Region

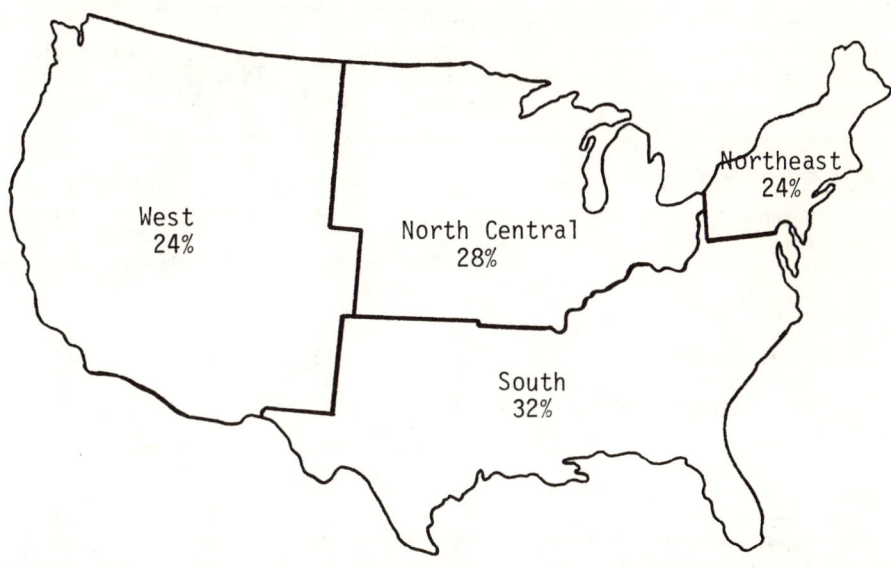

previous year (1981). Again, we asked: "Within the past year have other members of your family contacted an Extension agent or used the services of Extension?" Approximately 9 percent indicated that other members of the family had used Extension. This gave an overall household use measure of 14 percent for 1981 (over 11 million households).

Regional Use Patterns

Upon inspection of regional variation for the forty-eight contiguous states, the South reports a slightly larger proportion of households who had *ever* used Extension (a high of 32 percent compared with 24 percent in the West and Northeast as reported in Figure 4.1). Nevertheless, no statistically significant differences were present between the four regions ($p \leq .05$). On an annual basis, the South also had a slightly higher proportion of use than the other regions; but, again, it was not statistically different.

Rural-Urban Use

Sixty-four percent of the users of Extension live in metropolitan counties (in a Standard Metropolitan Statistical Area). Therefore, contrary

TABLE 4.2
Geographical Distribution of Users and Nonusers

	Ever Used		Previous Year(1981)	
	Nonusers (N=742)	Users (N=278)	Nonusers (N=883)	Users (N=143)
	%	%	%	%
SMSA County Classification				
Nonmetro	19	36	21	41
Metro	81	64	79	59
	$x^2=33.5*$		$x^2=27.9*$	
	Gamma=-.43		Gamma=-.46	
Current Residence				
Farm	4	10	4	13
Rural nonfarm	13	23	15	23
Town (less than 50,000)	28	30	29	28
City (50,000 or more)	55	37	52	36
	$x^2=40.9*$		$x^2=32.5*$	
	Gamma=-.33		Gamma=-.33	
Where Raised				
Farm	15	27	17	26
Rural nonfarm	16	21	17	20
Town (less than 50,000)	26	28	26	28
City (50,000 or more)	43	24	40	26
	$x^2=38.1*$		$x^2=12.7*$	
	Gamma=-.32		Gamma=-.23	
Farm Occupation				
Farmer	4	16	5	21
Nonfarmer	96	84	95	79
	$x^2=37.7*$		$x^2=46.4*$	
	Gamma=.60		Gamma=.64	

*$p \leq .01$.

to popular belief, the majority of Extension clientele are metro residents, with more than two-thirds living in towns and cities (see Table 4.2). This finding may come as a surprise to some individuals; however, it must be understood in the proper context. In sheer numbers metro users of Extension outnumber nonmetro clientele almost two to one, but one must remember that 73 percent of the U.S. population live in metro counties (U.S. Census, 1980). At the same time, Extension continues to serve a larger *proportion* of nonmetro residents than it does metro residents (42 percent versus 23 percent, not reported in table).

A similar finding is apparent for size of community of residence. Two-thirds of Extension users live in cities or towns; whereas, proportionally, more people on farms and in small towns use Extension than people in large cities. Also, proportionally, more farmers use Extension services than nonfarmers (57 percent and 25 percent, respectively).

The type of locality where people said they were raised shows a rural orientation among Extension users. People raised on farms or in small towns who now live in larger cities use the Extension service more extensively than people who were raised and currently live in large cities or metro areas. Thirty-five percent of urban users were raised in rural areas (compared with 18 percent of nonusers). Like a person's rural origin, previous 4-H involvement also influences present use patterns. It seems that those persons who were 4-H members as youth (26 percent of all users) are more likely to use Extension today and use it more frequently.

Therefore, three points can be derived from inspection of rural and urban use patterns. First, in sheer numbers, the majority of Extension's clientele reside in metro areas. Second, Extension is serving a greater proportion of rural people than urban. Third, many present-day clientele have roots in rural areas or small towns or were 4-Hers as youth.

Equity in Service Delivery

A typical equal opportunity statement for Extension reads: Extension is authorized to provide research, educational information, and other services only to individuals and institutions that function without regard to race, color, sex, age, or national origin. A question in the national survey explored this issue directly. Individuals were asked "Have you ever felt that you were discriminated against by Extension?" Only five individuals said yes. When these five respondents were asked "how" they were discriminated against, most answers were elusive or of questionable relevance to the equity issues. One individual said he "signed up for feed during a drought and didn't get any." This statement is obviously in reference to another agency's program. Another said that Extension "didn't let me know about their services through advertisement." A third stated: "I am a homosexual. They gave me little information when I asked for it." A fourth said that she was discriminated against because of her race. The fifth would not specify "how." Only the latter statements appear to manifest possible elements of discrimination according to the criteria specified.

Information was also gathered on the race, and/or ethnic origin, sex, and age of all respondents (Table 4.3). These equity criteria were then related to the use of Extension. No significant variation was apparent concerning sex or age status in use patterns for the previous year. However, significant age variation is apparent between users and nonusers for those who had ever used Extension. Middle-aged groups (30 to 64)

TABLE 4.3
Equity for Users and Nonusers

Equity Issues	Ever Used		Previous Year(1981)	
	Nonusers (N=742)	Users (N=278)	Nonusers (N=883)	Users (N=143)
	%	%	%	%
Race				
White	82	94	84	92
Black	11	4	10	3
Hispanic (Puerto Rican, Mexican, etc.)	5	1	4	2
American Indian	1	1	1	2
Others	2	1	1	1
	$x^2=24.4*$		$x^2=9.8*$	
Sex				
Female	57	62	58	58
Male	43	38	42	42
	$x^2=2.14$ Gamma=-.11		$x^2=0.00$ Gamma=-.00	
Age				
Less than 30	37	24	34	26
30-39	20	28	21	31
40-64	31	40	33	36
65 and over	12	8	12	7
	$x^2=24.1*$ Gamma=.12		$x^2=10.6$ Gamma=.04	

*$p \leq .01$.

were overrepresented among users, while younger adults (18 to 29) and older adults (65 and over) were underrepresented. While the same trend was apparent for age variation in use patterns during 1981, this relationship was nonsignificant.

Rather substantial variation was noted in use patterns of racial and/or ethnic groups. Clearly, whites use Extension more than nonwhites. Of all racial/ethnic groups studied, blacks were the most underrepresented. This applied to both total previous use and use for one year. However, only one nonwhite voiced concern about discrimination when directly asked.

In summary, no substantial differences were apparent between patterns of use for persons of different ages or sexes during 1981. On the basis of race, very few respondents raised possible legitimate discrimination issues when asked. Given the fact that respondents were directly queried about discrimination (with subsequent probing), this seemed to reveal little felt discrimination by the public. It is only on racial/ethnic iden-

TABLE 4.4
Socioeconomic Characteristics of Users and Nonusers

	Ever Used		Previous Year(1981)	
	Nonusers (N=742)	Users (N=278)	Nonusers (N=883)	Users (N=143)
	%	%	%	%
Family Income (before taxes)				
Less than $5,000	10	3	9	3
$5,000-$9,999	16	10	15	9
$10,000-$19,999	27	27	27	25
$20,000-$29,999	25	26	25	29
$30,000-$39,999	12	16	13	16
$40,000-$49,999	4	8	5	7
$50,000 and over	6	9	6	11
	$x^2=28.7*$		$x^2=15.0$	
	Gamma=.24		Gamma=-.23	
Education				
Grade school	8	5	9	2
High school	54	45	53	41
College	32	32	31	34
Graduate degree	6	18	7	23
	$x^2=39.4*$		$x^2=47.0*$	
	Gamma=.27		Gamma=.40	
Marital Status				
Married	56	73	59	74
Separated	3	1	3	2
Divorced	9	7	9	4
Widowed	8	5	7	6
Never married	24	14	22	15
	$x^2=26.2*$		$x^2=12.72$	
Employment Status				
Employed	60	69	61	71
Unemployed	9	3	9	2
Retired	12	10	12	8
Homemaker	14	15	13	18
Student	5	3	5	1
	$x^2=16.5*$		$x^2=17.0*$	
Residence				
Own home	58	79	61	82
Rent	36	19	34	16
Live with relatives	5	2	4	2
Other	1	0	1	0
	$x^2=39.0*$		$x^2=22.97*$	

*$p \leq .01$.

tification that variation is apparent. While very few overt feelings of discrimination or inequity were reported by the respondents, clearly, Extension is serving a lower proportion of nonwhites.

Profile of Users

Extension clientele are predominately middle class. They are middle to upper income, high school and college educated, white, married, employed, and homeowners. The study of use patterns indicates an underrepresentation among Extension clientele of: the poor; single, divorced, separated/widowed persons; those with less educational attainment; the unemployed, retired, or students; and renters (Table 4.4). The underrepresentation of nonwhites has already been noted. In short, Extension seems to reach the vast white, stable, middle segment of Americans.

Clientele and Political Activity

Extension's clientele are active in the political arena. Users of Extension services are significantly more likely to vote and to contribute to political candidates than are nonusers (Table 4.5). For example, over one-third of Extension clientele contributed time and/or money to political candidates during the 1980 elections, while only one out of five nonusers contributed. However, the traditional image of Extension clientele as conservative does not hold. No significant variation was found in the political ideology of users and nonusers. This higher level of political activity among Extension users is consistent with the political patterns of the middle to upper-middle portion of the population in general (Vanfossen, 1979).

Use of the Four Program Areas

To understand more fully use patterns of Extension services, household use for the year 1981 was studied for each of the four program areas (Figure 4.2). Agricultural programs had the highest level of use of the four programs. Approximately 62 percent of all respondents who indicated use during the previous year had received assistance in agriculture. This figure represents about 9 percent of the total adult population. Home economics programs were second with slightly over 40 percent reporting use in 1981. Community development and 4-H programs had between 20 and 30 percent using these program areas. The 4-H program is likely to reflect an underreporting since respondents were adults, while the 4-H program is directed at youth.

TABLE 4.5
Political Involvement and Ideology of Users and Nonusers

Political Orientation	Ever Used		Previous Year(1981)	
	Nonusers (N=742)	Users (N=278)	Nonusers (N=883)	Users (N=143)
	%	%	%	%
Vote, 1980				
No	32	14	29	16
Yes	68	86	71	84
	x^2=31.7*		x^2=11.03	
	Gamma=.48		Gamma=.38	
Contribute				
No	80	68	79	64
Yes	20	32	21	36
	x^2=11.03		x^2=14.6*	
	Gamma=.38		Gamma=.35	
Political Orientation				
Conservative	34	38	34	39
Middle-of-the-road	42	43	43	41
Liberal	24	19	23	19
	x^2=3.82		x^2=1.6	
	Gamma=-.11		Gamma=-.10	

*$p \leq .01$.

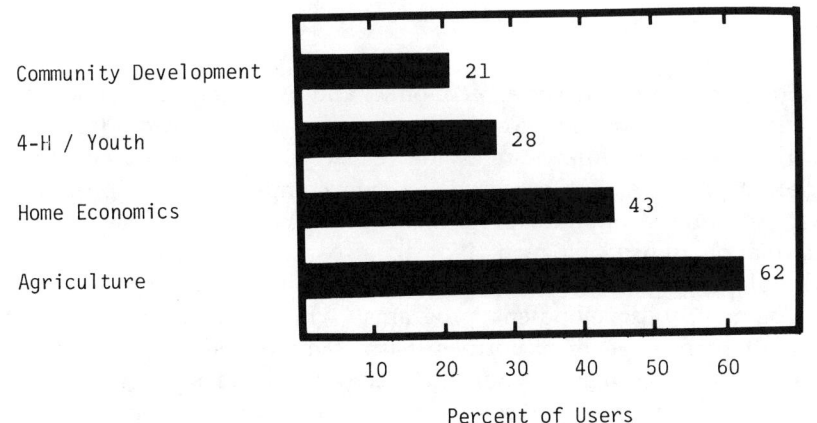

FIGURE 4.2
Household Use of Four Program Areas, 1981

Community Development 21
4-H / Youth 28
Home Economics 43
Agriculture 62

Percent of Users

TABLE 4.6
Characteristics of Users of the Four Program Areas in 1981

Select Characteristics of Users	Agriculture (N=88)	Home Economics (N=61)	4-H (N=41)	Community Development (N=30)
	%	%	%	%
Sex				
Female	49	72	60	57
Male	51	28	40	43
Race				
White	93	86	85	79
Black	3	5	5	14
Other	3	9	10	7
Occupation				
Farmer	30	18	28	13
Nonfarmer	70	82	72	87
Current Residence				
Farm	18	10	18	0
Rural nonfarm	23	28	21	20
Town	21	32	33	30
City (50,000+)	38	30	28	50
Income				
Less than $10,000	7	14	8	10
$10,000-$19,999	27	34	37	38
$20,000-$29,999	26	27	13	28
$30,000 or more	39	25	42	24
Education				
Grade school	1	2	3	0
High school	41	48	28	40
College	31	35	38	33
Graduate school	26	15	31	27

Comparisons of the social, economic, and demographic characteristics of users of the four program areas were also examined (Table 4.6).[2] Two findings are interesting with respect to the sex of clientele of different programs. As expected, women comprise the majority of the users of home economics programs, but a quarter of home economics information is used by men. Women also make extensive use of agricultural programs. More black and other minority groups were users of community development programs, while agriculture served the smallest percentage of minorities. Farm families used agricultural programs and 4-H programs more than they used other program areas. It is noteworthy that 70 percent of users of agricultural programs are not farmers. In fact, 60 percent of the clients of agricultural programs live in towns and cities. As many as 80 percent of community development program users reside in urban areas. Agricultural and 4-H programs tend to serve persons of higher income and educational levels. Forty percent of agricultural and 4-H clients have an income of $30,000 or greater. Furthermore, 69 percent of 4-H users (as well as a majority of the other three program areas) are college educated.

Frequency of Use

The Kentucky study of Extension users inquired as to the frequency of clientele use. Though not reported as a rate of use (i.e. number of times per month or year), Table 4.7 provides an indication as to whether clientele use Extension services rarely, occasionally, or frequently. Frequent users describe persons who have established a consistent pattern of use over time and turn to Extension for assistance on a regular basis. Occasional users draw upon the services of Extension on an as-needed basis with contact being irregular. In contrast, some persons could be classified as rare users in that they use Extension very infrequently, maybe once or twice a year.

Over half of Extension users (58 percent) consider themselves to be occasional or frequent users, while 42 percent use the service only rarely. Frequent users comprise a small proportion of all users (12 percent). They could be described as the loyal followers of Extension and would be expected to be the organization's strongest supporters. As one might expect, the most frequent users of Extension programs are farm residents. On the basis of both acreage and sales, farmers who operate large-scale farms are more frequent users than are operators of small and medium-sized farms.

Methods of Use

In addition to the question who uses Extension, respondents were also asked how they used or contacted (were contacted by) Extension. Three common methods utilized by Extension were incorporated in the questions. As can be seen in Table 4.8, almost all of the users (99 percent) had received some printed material from Extension. Over 90 percent had listened to a radio program or watched a television program conducted by Extension personnel. In contrast, only 39 percent of users (5 percent of the U.S. population) had attended an Extension workshop or meeting in the past year. However, on the national level, this percentage translates into 4.3 million households in attendance at such workshops

TABLE 4.7
Frequency of Use by Kentucky Extension Clientele

	Frequency of Use		
Characteristics	Rarely	Occassionally	Frequently
	%	%	%
Total Population (3,289)	42	46	12
Place of Residence			
Farm	33	52	15
Rural nonfarm	49	42	9
Town	47	43	10
City	59	39	2
	$x^2=992.7$,	Gamma = -.40	
All Farms (N=1,409)	33	52	15
Farm Sales			
Less than $10,000	37	50	13
$10,000 - $19,999	35	46	19
$20,000 - $39,999	27	54	17
$40,000 or more	18	53	29
	$x^2= 33.0$,	Gamma = .21	
Farm Size (acres)			
Less than 100	38	49	13
100 - 259	32	53	15
260 - 499	22	58	20
500 or more	24	49	27
	$x^2= 32.0$,	Gamma = .19	

and meetings. Overall, most Extension contact seems to come through publications or mass media rather than through group meetings.

Dilemmas of Whom to Serve

More than one out of every four households in the U.S. have used the services of Extension, with about half of those using it in the past year. Two-thirds of these households are located in urban areas. Therefore, even though Extension has a history of service to rural areas, most of today's clients do not live on the rural farmsteads of past generations.

TABLE 4.8
Methods of Communication Utilized by Clientele

	Households Utilizing this Method	
Methods	Percent of Total (N=1,028)	Percent of Users (N=143)
Written Material (bulletins, newsletters, or publications)	14	99
Radio or TV	13	94
Meeting or Workshop	5	39

Society has changed, people's place of residence has changed, and Extension has changed as well.

For all the diversity of state Extension services, the pattern of use is surprisingly uniform throughout the nation. Although slightly higher in the South, the extent of use is almost the same in all four regions of the country.

The national survey reports a greater number of urban clientele; however, Extension continues to serve a greater proportion of rural and farm residents. This finding suggests that the agency has not abandoned these traditional clientele in its quest to serve urbanites but, rather, continues to focus considerable attention on this rural audience. The issue is whether the organization has drifted too far away from its target audience as traditionally defined or whether it has too long hung onto a rural and farm clientele group that is diminishing in numbers and influence.

Almost no incidences of discrimination were uncovered in the national survey. However, Extension continues to reach a lower proportion of nonwhites than is present in the total population. This disparity suggests that Extension has generally not been guilty of treating participants differently based on such factors as race, age, and sex, but indirectly some individuals and groups may be excluded from participation because programs do not relate to their needs. Rather than spending its time stressing the equal-opportunity nature of its programs, maybe Extension ought to be more concerned with the appropriateness of program content and methods for reaching particular target audiences. Extension's heavy reliance on printed materials may be a factor discouraging participation by individuals with a low level of educational achievement. And, in terms of content, the community development program seems to have greater appeal to minority groups than do the other program areas.

The statement that Extension predominantly serves a middle-class clientele is largely correct. However, the description of Extension clientele approximates closely the U.S. adult population. As a voluntary educational program, Extension attracts the large middle-class majority.

This chapter answers the question of whom Extension is reaching with its programs. However, it does not address the larger issue of whom Extension *should* be serving. That question is left to policymakers.

Notes

1. The extent of use reported in these studies would be expected to vary somewhat due to differences in wording of questions, variation in the time periods examined, and differences in the structural definition of Extension (University Extension versus Cooperative Extension).

2. Caution must be exercised when interpreting differences in personal characteristics because of the small Ns in some categories, respondents' reported characteristics were used to represent household characteristics, and respondents could report use of more than one program area. Tests of differences among users and nonusers are included in Appendix B, Table B.3.

5
Some Clients Are Satisfied and Some Are Not

The effective organization is the organization that best serves those who perceive it as a means to their ends.
—Cummings, 1977:61

After it is all said and done, programs may be properly designed and efficiently run, but unless they respond to the demands and expectations of the intended recipients of the programs, they can be seen as ineffective and a waste of taxpayers' money. The bottom line is, does the program really make a difference in the lives of the people? Do changes result? What do the recipients of the program think of the service they receive? Does it serve their needs? Effectiveness from this perspective, then, is the degree to which the organization is meeting the needs of constituents.

Subjective and Objective Performance Measures

Extension and other public agencies increasingly are required to focus on the outcomes or "effects" of their programs. However, differences of opinion surface when one attempts to determine the technique for determining program impacts. Probably the most commonly used method has been to compare outcomes with statements of intended purpose—the goal approach. Effectiveness, in terms of this approach, is defined as goal attainment. The problem is that formal goals primarily reflect the thinking of organizational members and are seldom widely known by constituents. Therefore if goals are to be utilized as criteria for determining effectiveness, what must be recognized is that these expressions represent a perspective from within the organization. The Extension

Management Information System is an example of the goal-oriented view.

The goal-approach to determining organizational effectiveness is generally seen as producing objective measures of output, while perception indicators are viewed as subjective in nature. Objective measures reflect quantifiable measures of service performance, while subjective indicators provide citizens' qualitative evaluations of services provided. For example, dropout rates and test scores would be considered objective performance measures for evaluating the educational achievement of students, while citizen opinions concerning the quality of education would be treated as subjective information.

Quality is in the eye of the beholder (Campbell and Converse, 1972). Perception measures of organizational performance provide citizens' qualitative evaluations of service provided. Increasingly, there has been a reliance upon subjective appraisals; however, some evaluators who have been accustomed to highly quantifiable "objective" indicators find subjective measures inadequate. Scott (1977), for example, sees subjective measures as useful when objective indicators are unavailable but still questions the validity of such measures. In contrast, proponents of clientele appraisals question whether assessments of organizational effectiveness can ever be purely objective in nature or whether organizationally constructed evaluation criteria can truly reflect the varied needs of the many segments of the target population. All measures of effectiveness are value laden and to seek true objective indicators of effectiveness "is to pursue an illusion" (Scott, 1977:69). Moreover, the goal approach to evaluation may not address the important issues as defined by the ultimate user of the service.

Client Satisfaction

Client satisfaction has come to be accepted as a legitimate measure of system output (Shin, 1977; Stagner, 1970). In the medical field, reputational measures of effectiveness have been employed with a high level of reliability (Georgopoulos and Mann, 1962; Georgopoulos and Tannenbaum, 1957). Other examples in the use of subjective assessments by clientele include nutritional education programs (Development Planning Associates, 1977), welfare programs (Rossi et al., 1974), police protection (Kelling et al., 1974), and other public services (Christenson et al., 1980; Katz et al., 1977). In short, client satisfaction is a commonly used qualitative indicator of public services (Andrews and Withey, 1976; Campbell et al., 1976; Marans and Rodgers, 1975).

Though consumer satisfaction with public agencies has become an accepted measure of organizational effectiveness, there remains the

TABLE 5.1
User Satisfaction with Government Service Agencies

Agency	Percent Satisfied
Social Security	88
Workmens Compensation	75
Job Training	74
Unemployment Compensation	71
Public Assistance	61
Employment Service	61
Hospital/Medical Care	58
Overall Average	69

Source: Katz et al., 1977: 63-64.

problem of determining an acceptable or appropriate level of satisfaction to expect. Traditionally, comparisons of levels of satisfaction for different social or economic groups have been employed. More recently, Katz et al. (1977) have suggested threshold levels for government agencies. Looking at a wide variety of government service agencies, they found that approximately two-thirds of users were satisfied with their experience with government agencies. Users of Social Security reported the highest level of satisfaction (88 percent), while the levels of satisfaction with public assistance, employment services, and medical care were somewhat lower (approximately 60 percent, Table 5.1). However, Katz and associates caution that such levels of satisfaction should not necessarily be accepted as the standards of excellence; some services may require much more ambitious standards of performance, while others would need to be lower.

In utilizing client satisfaction as an indicator of organizational effectiveness, we are assuming that as performance changes, so do levels of satisfaction. In reality, improved performance may lead to qualitative differences in response but not necessarily quantitative changes (Maslow, 1971). The number of complaints about an organization may remain unchanged but merely vary in content (Weick, 1977). In fact, the term "hedonic treadmill" has been used to describe the way in which

FIGURE 5.1
Satisfaction with Extension by Region of the U.S.

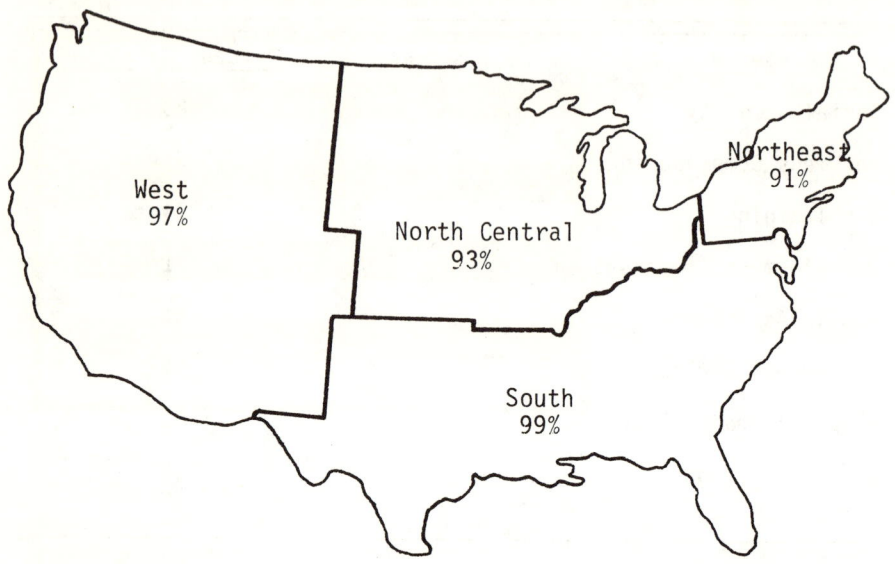

expectations, wants, and desires escalate rather than decline as an organization becomes more effective (Brickman and Campbell, 1971).

It has also been suggested that the level of expressed satisfaction is related to a person's socioeconomic status (Katz et al., 1977); that higher socioeconomic status persons are more critical of services than those of a lower level. Katz et al. (1977:194), therefore, conclude that, "We thus may want to discount some of the satisfaction expressed by the poorly educated and be cautious about accepting at face value all of the criticisms of the well educated."

Satisfaction: A National Perspective

In the national sample, more than nine out of ten persons who had *used* the Cooperative Extension Service were satisfied with the services they received. This level of satisfaction is uniformly high across the nation, though it is slightly higher in the South and the West (Figure 5.1). Therefore, in response to the question, "Are Extension clientele satisfied?" the answer is an overwhelming *yes*.

Most any agency would be pleased with this level of client satisfaction in that it is substantially higher than the 69 percent satisfaction that Katz et al. (1977) found for most government agencies. Satisfaction is

FIGURE 5.2
Satisfaction with Extension Programs by Users of the Service

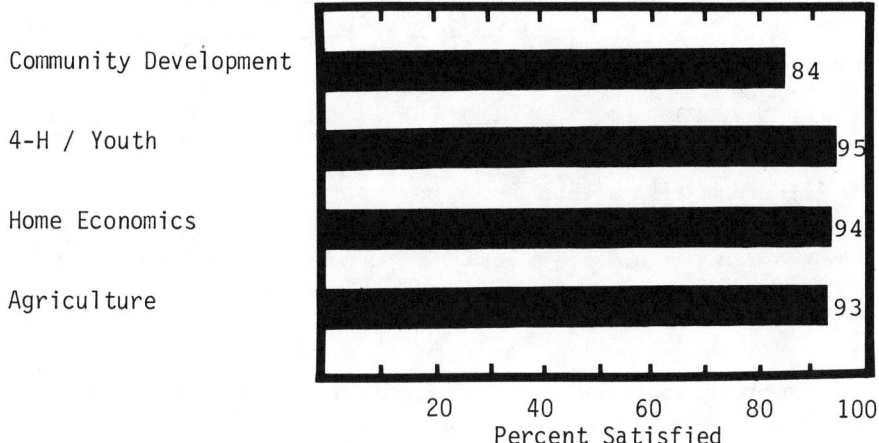

also very high for all of the four program areas. The only one that is less than 90 percent satisfied is community development at 84 percent (Figure 5.2).

Upon examining the satisfaction level by the characteristics of the individuals responding, *no* significant differences were found on the variables age, sex, income, education, place of residence, race, and occupation (Appendix B, Table B.4). Therefore, not only is the overall level of satisfaction very high, but it is also uniformly high among all segments of the population whether they be rich or poor, urban or rural, black or white.

Since respondents were also asked to rate each program area, it is possible to relate the background characteristics of clients to the program satisfaction ratings. Sex, race, income, education, place of residence, and occupation make no difference in the respondents' assessment of satisfaction of the four program areas (Appendix B, Table B.5). Age proved to be significant only with the home economics and community development programs, in which the young and the old were less satisfied.

The national study asked the satisfaction question of only those persons indicating that a member of the household had used the services of Extension. In other words, nonusers were not asked to rate a service with which they were not expected to be familiar. Conceptually, this makes sense, but one has to question whether this level of expressed satisfaction is artificially high because the procedures excluded those who are nonusers and possibly dissatisfied. Though not directly served

TABLE 5.2
Satisfaction with Extension in Kentucky

	Dissatisfied %	Unsure %	Satisfied %	Total %
Satisfaction[a]				
Users (N=3,287)	10	19	71	100
Nonusers (N=6,458)	14	46	40	100
Satisfaction (recalculated without unsure category)				
Users	12		88	100
Nonusers	26		74	100

[a] $p \leq .05$.

by programs of Extension, it is anticipated that a considerable number of nonusers have formed an opinion of Extension.

As the U.S. General Accounting Office (1978) suggests, not only should beneficiaries of programs be queried, but also the cost-bearers. Therefore, there is a need to examine the attitudes of nonusers as well as users of the service. A state study similar to the national one was utilized for this purpose. The state study contains sufficient sample size to enable a detailed analysis of subgroups of users and nonusers not possible in the national study.

Satisfaction: A State Perspective

Information comparable to that contained in the national study was also collected within a single state. A mailed questionnaire was sent to a sample of registered voters in Kentucky in 1979, from which 11,015 persons (69.1 percent) responded. In this state survey, a question concerning satisfaction with Extension was asked of all respondents whether they used the service or not. It also allowed persons the option of choosing an unsure category if they felt they lacked adequate information from which to make a decision. As seen in Table 5.2, more than seven out of ten persons who had used the service indicated satisfaction with Kentucky Extension. However, if the unsure category is removed to make it comparable to the national study, the proportion of satisfied users is 88 percent, or about the same as reported for the U.S. as a whole.

In the nonuser category, slightly more than half (54 percent) of the respondents felt they had enough information about Extension to judge whether they were satisfied or not. Among nonusers, 40 percent were satisfied (compared with 71 percent for users). These results confirm earlier observations that many nonusers do have opinions about Extension, and when compared with users they are less likely to be satisfied. Again, if the unsure category is removed, the recalculated satisfied percentage for nonusers is 74 (compared with 88). This finding from the Kentucky study confirms the speculation that users are highly satisfied, whereas nonusers are somewhat less satisfied. However, one should note that the level of expressed *dissatisfaction* is about the same for both users and nonusers (10 percent compared with 14 percent). Thus, the difference in the satisfaction ratings is related to how many persons lack sufficient information upon which to judge the merits of Extension (the number in the unsure category). Overall, there exists a modest relationship between use and satisfaction ($r = .27$).

Users and Nonusers

The Kentucky study allows for an analysis of users' and nonusers' assessment of Extension. In general, one can conclude that the results of the state study follow the patterns of the national study. However, with a considerably larger sample size, some differences prove to be statistically significant in the state data that were not so in the national findings. Also, differences would be expected to reflect the unique nature of Kentucky's program. For example, in the Kentucky study, older persons were found to be more satisfied with Extension than were young people, and farmers expressed more satisfaction than did rural nonfarm, town, and city residents. None of these differences is significant in the national data set.

Users

There are significant differences in the satisfaction level of clientele according to the person's age, family income, education, place of residence and farm size. Satisfaction increases with age, from 63 percent among the youngest group to almost 80 percent for the elderly (Table 5.3). The highest income category registered the greatest satisfaction with Extension, while satisfaction was about the same for other income groups. Satisfaction is higher for those persons with more education, especially those with a graduate degree. Farm users are more satisfied than nonfarmers, particularly if they operate farms of over 260 acres.

There are also certain groups of users who are more dissatisfied with Extension. From the above explanation, one would expect, and the findings confirm, greater dissatisfaction among young people, low-income persons, individuals with a lower level of education, and operators of small farms (under 260 acres). Contrary to expectations, rural nonfarm

TABLE 5.3
Satisfaction of Users of Kentucky Extension by Personal Characteristics
(N=3,287)

	Dissatisfied %	Unsure %	Satisfied %	χ^2
Sex				
Male	11	18	71	
Female	9	19	72	2.10
Age				
18 - 29	12	25	63	
30 - 39	8	21	71	
40 - 64	10	16	74	
65 and over	9	12	79	51.16*
Race				
White	10	18	72	
Nonwhite	13	18	69	0.77
Income				
Less than $10,000	13	19	68	
$10,000 - $19,999	9	18	73	
$20,000 - $29,999	7	21	72	
$30,000 and over	7	14	79	29.75*
Education				
Grade school	14	20	66	
High school	10	19	71	
College	7	19	74	
Graduate degree	8	15	77	26.36*
Place of Residence				
Farm	10	14	76	
Rural nonfarm	11	23	66	
Town	7	19	74	
City	9	21	70	42.13*
Farm Size (acres)				
Less than 50	11	16	73	
50 - 179	12	14	74	
180 - 259	11	16	73	
260 - 499	7	11	82	
500 or more	4	14	82	11.93*
Farm Sales				
Less than $10,000	11	16	73	
$10,000 - $19,999	3	11	86	
$20,000 - $39,999	10	14	75	
$40,000 or more	5	10	85	20.68*

*$p \leq .05$.

residents are more dissatisfied than town and urban residents. Though it is somewhat surprising that rural nonfarm clientele are less satisfied (and more dissatisfied) with Extension than are their urban neighbors, it is clear that the lower level of satisfaction is not coming from farmers but rather from other rural residents.

There is no difference in the satisfaction level for two equal-opportunity criteria—sex and race. Males and females, whites and nonwhites are equally satisfied with Extension.

Nonusers

In many cases, nonusers responded differently than users on the satisfaction question. Interestingly, these nonusers often hold more extreme positions than do users. In fact, certain groups of nonusers are likely to be both more satisfied and more dissatisfied with Extension. Moreover, these disparities occur within specific segments of the population. Persons of low socioeconomic status, as evidenced by income and education, are more satisfied *and* more dissatisfied with Extension than are persons of higher socioeconomic position (Table 5.4). Likewise, rural farm and nonfarm residents hold similar extreme views. These findings reflect substantial differences of opinion on Extension.

Among nonusers, more farmers and rural nonfarm people hold a positive opinion about Extension (40 to 50 percent compared with about 30 percent for town and city residents). There seems to be a predilection toward satisfaction with Extension among farmers and rural people even if they have not used the service. Urban areas, unlike rural areas, have not had a long history of Extension involvement from which to develop such a positive image of the organization. Thus, mere exposure to the programs of Extension in urban areas becomes very important in generating satisfaction with the organization.

As with users, there is a clearly defined pattern of greater satisfaction among nonusers as age increases. Thirty percent of those persons in the youngest age category express satisfaction with Extension, while over half of those over 65 years old are satisfied. There are only small differences in the satisfaction level according to the race of the respondent.

How Frequency of Use Affects Satisfaction

The more frequently people use the services of Extension, the more satisfied they are with it. The relationship between frequency of use and satisfaction is strong (gamma = .48). Four out of five people who are frequent users report being satisfied, compared with about half of the occasional users and nonusers (Table 5.5).

TABLE 5.4
Satisfaction of Nonusers of Kentucky Extension by Personal Characteristics
(N=6,458)

	Dissatisfied %	Unsure %	Satisfied %	x^2
Sex				
Male	14	45	41	
Female	13	47	40	7.12*
Age				
18 - 29	12	57	31	
30 - 39	13	51	36	
40 - 64	14	41	45	
65 and over	15	32	53	191.36*
Income				
Less than $10,000	15	42	43	
$10,000 - $19,999	13	49	38	
$20,000 - $29,999	12	49	39	
$30,000 and over	11	52	37	38.62*
Education				
Grade school	18	44	48	
High school	12	48	40	
College	12	52	36	
Graduate degree	13	49	38	108.58*
Place of Residence				
Farm	17	32	51	
Rural nonfarm	14	45	41	
Town	11	52	37	
City	12	58	30	141.37*
Farm Size (acres)				
Less than 50	17	33	50	
50 - 179	16	25	59	
180 - 259	19	30	51	
260 - 499	16	34	50	
500 or more	20	30	50	10.99
Farm Sales				
Less than $10,000	16	34	50	
$10,000 - $19,999	15	24	61	
$20,000 - $39,999	14	30	56	
$40,000 or more	16	37	47	4.72

*$p \leq .05$.

TABLE 5.5
Satisfaction with Kentucky Extension by Frequency of Use

Number of times used	Dissatisfied %	Unsure %	Satisfied %
None	14	46	40
Occasional	12	28	60
Frequent	8	11	81

x^2 = 1072.4; Gamma = .48.

This finding carries the discussion one step further in that it relates the intensity of involvement of clientele to their assessment of the organization. It allows us to differentiate between the occasional and frequent user rather than just the user and nonuser. The level of satisfaction is considerably greater among persons who make repeated use of the service over those who only occasionally use it.

Issues of Service Assessment

Users Are Satisfied

Is Extension effective? Clientele think so. Once people use Extension, they like the services they receive. On the national level, 95 percent of users are satisfied with Extension. This figure is much higher than that found in the assessment of organizational effectiveness of other public agencies (Katz et al., 1977). All types of users of Extension are highly satisfied irrespective of their age, sex, income level, educational level, race, occupation, and place of residence. Finally, the more frequent the use, the higher the level of satisfaction.

The General Public Is Unsure About Extension

Is Extension seen as effective by the general public? What is the perception of the three-fourths of the population who are not Extension users? Four out of ten nonusers have a positive opinion of Extension, while 14 percent are dissatisfied. And, as one might expect, about half of the nonusers have not formed an opinion of Extension, either positive or negative. It is important to recognize that this large "unsure" group represents about one-third of the total population and, therefore, could have an important influence on the future of Extension. They are cost-bearers and have a stake in the allocation of public resources.

Upper Middle Class Is More Satisfied

High socioeconomic status (SES) individuals are more satisfied with Extension, and the greatest dissatisfaction is among persons of low income and educational levels. This is contrary to the findings of the Katz study of other public service agencies (1977). Either the Katz finding does not hold for an educational organization like Extension, or Extension is actually serving high SES persons better. It has been demonstrated in Chapter 4 that Extension tends to serve a greater proportion of medium and high SES individuals than it does low SES persons. As a result, one could expect these higher income and better educated persons to be more satisfied with the service they are receiving. Also, the Katz study concentrated on service-type agencies, while Extension is educational in nature. A voluntary educational program is more likely to appeal to persons who can understand and utilize the informational content of the program. Though Extension attempts to direct its programs at all educational and income levels, the fact remains that, in order to take advantage of the bulk of Extension materials and programs, at least a minimum level of educational proficiency and economic flexibility is required. Therefore, the pattern of response found is not totally unexpected, but it does suggest that Extension ought to assess the appropriateness of its programs for this lower socioeconomic audience.

Small Farm and Rural Nonfarm Residents Are Less Satisfied

Farmers are very satisfied with Extension, especially those operating large-sized farms. Evidently, Extension has served their needs well. However, Extension may need to examine its service to operators of small farms. They are substantially less pleased with the service they have been receiving.

Rural nonfarm residents represent a group of persons living in rural areas but not on farms. These persons have been identified as appropriate clientele of Extension since the organization's inception and continue to be seen so today. Nevertheless, these rural nonfarm clientele are less satisfied with Extension than farmers, small town residents, or residents of large cities. Extension needs to examine why this occurs. Are the needs of this group being inadequately addressed, or are these persons just more critical of Extension? On the one hand, one might argue that, with population turnaround and the ensuing movement of people to rural areas as a place to live, Extension has focused insufficient attention on the unique needs of this new type of rural resident. On the other hand, there is sufficient evidence in the present study (data not presented) to suggest that this group of people is not only more critical of Extension but of most other public services as well. No doubt at least some of this perception accurately reflects a general lower quality of services available in some rural areas as well as the fact that these persons' expectations for their public services may be greater than what are available. Therefore, this group's assessment of the quality of services of a single agency like the Extension Service may be less a reflection of that agency's performance than it is an expression of a general level of dissatisfaction among these residents.

The Elderly Are Well Served

The state study clearly demonstrates that older persons are more satisfied with Extension than are young people. This finding holds for both users and nonusers. The important question is why are young persons less satisfied? Is the organization doing a poorer job of serving this young group? Or do older persons consistently give a higher rating to all services?

If, in fact, younger persons feel that Extension is not adequately responding to their needs, then program staff ought to be concerned with the nature of programs and their intended audiences. Thought may need to be given to developing programs that more satisfactorily speak to the problems of persons and their families in the 18 to 30 age range. It is true that the elderly tend to rate some services higher than do other groups; however, they also are quite critical of certain services that in their opinion fail to meet their needs. Therefore, one cannot assume that Extension's higher satisfaction rating by the elderly can be attributed solely to a more charitable overall assessment by older people. Extension needs to review the nature of its service to different age groups not only because of its implications for present programming but also because young people represent the clientele of the future.

6
Support for Extension

> *The ultimate decision to give or withhold the needed organizational inputs lies in the environment, . . . the larger social environment in this way holds the power of life and death over every organization.*
> —Katz and Kahn, 1978:238

An organization must have resources in order to survive and grow. Without adequate support, it will wither and die. It could be argued that organizational survival is the ultimate criterion of success, because, without sufficient resources for maintaining the organization, other effectiveness criteria are meaningless.

The effectiveness of an organization is measured not only in terms of the accomplishment of programming goals but also "in the acquisition of scarce and valued resources" (Yuchtman and Seashore, 1967:898). In this approach there is a recognition that, in addition to programming goals, an organization must also be concerned with resource procurement (Thompson, 1967).

With increased size and complexity of organizations comes a separation of consumption and control. The users of services often do not make the decisions that affect the nature of the delivery organization. Like corporations that have a separation between the role of the consumer and the owners of the company, public service organizations have a separation between the recipients of the services and policymakers who control the allocation of resources (Etzioni, 1964). The greatest separation between consumption and control is generally found in large, public monopolies, while more consumer control is found in small, personal-service type agencies. In the private sector, consumer influence is exerted through the choice of purchasing or not purchasing a product or service. In the public sector, the linkage is not so direct. The activities of public agencies are financed through taxation with elected and appointed representatives making decisions on resource allocation. The public elects these officials and influences their decisions, but the right to vote for or against a political candidate is often far removed from a person's feelings about a specific program or agency. Though the ballot box is one method of consumer control, it is a very indirect one. Also, not all consumers are in a position to exercise control in this manner. For example, 4-Hers are not old enough to vote.

Support of public agencies is crucial to their success. The formal procedure for consumers to influence the allocation of resources is through the use of the political process; however, a number of factors unique to an organization and/or its programs influence the likelihood of public support for an organization.

Clientele Involvement

Extension encourages direct involvement of clientele in the process of planning and carrying out programs. Citizens' advisory councils and committees are active in the formulation of Extension programs at the local, state, and federal levels. Through this process of participation, people come to know the organization. Organizations that have a high degree of participation are more likely to be seen as effective by the intended recipients of the programs (Price, 1968). There develop extensive networks of primary relationships through personal, face-to-face interaction and these primary relationships result in more extensive support for the organization. One of the expectations of these "friends of the organization" is that they will render assistance in times of need. The nature of the assistance is general support of organizational efforts.

In an organization like Extension that promotes extensive client involvement, it would be anticipated that there would exist a general feeling of goodwill among its constituents. The nature of the support would not be specified but would exist as a diffuse obligation or commitment to the agency and its programs (Price, 1968). In all likelihood, this positive sentiment would lay dormant until the organization is threatened. In which case, support could be mobilized for specific purposes.

Autonomy of Extension

Extension is a locally based organization that attempts to be responsive to the needs of local citizens. Though there is state and federal input in the determination of program priorities, the emphasis is on the local control of organizational decisions. As one observer concluded (DeMarco, 1980:3), "Extension is not generally viewed as part of a government agency at all. Many of the people . . . [are] only vaguely aware of its Federal and State ties."

Greater autonomy allows an organization more adaptability and flexibility in dealing with local problems. A unique feature of Extension, in comparison with most other government agencies, is its high degree of decentralization in decision making. This decentralized decision-making structure is an important factor influencing the extent of local support. The greater the autonomy of an organization, the more likely local citizens are to see it as effective and be willing to support it (Price, 1968).

Local Interests

Support for an organization depends upon the extent to which the organization is based on principles that are widely accepted by local citizens. In other words, are the ideas and practices of the organization consistent with those of the general public? In some cases, the organization may be promoting ideas that are contrary to the values and norms of the local culture. Such organizations are not likely to garner widespread support. In fact, residents may go out of their way to criticize and attack an organization that is seen as incompatible with their way of life.

Extension has a strong local identity. Agency personnel are local residents. They live and work in the county. And while Extension agents are respected for their technical expertise, they are also treated as friends and neighbors. The content of Extension's educational program is largely determined through the involvement of people through the program-planning process, and advisory groups have a direct say in matters concerning personnel, budgets, and programs. After visiting ten state Extension Services as part of the 1977 Congressionally mandated evaluation of Extension, DeMarco (1980:4) commented that, "the atmosphere in which Extension workers dealt with clients was one of mutual respect" and that "Extension workers, unlike many others on the public payroll seem to understand that their jobs are to deliver a service to the client's satisfaction and the degree to which Extension workers are willing to get out of their offices and work with clients is the degree to which they are effective in maintaining support."

Positive public opinion about an organization results in a perception of greater organizational effectiveness and support. The greater the extent to which the efforts of the organization are viewed as consistent with the needs of the public, the more willing people are to support the organization (Meier and Browne, 1983). People tend to support what they feel is useful. A favorable public image makes securing resources from funding sources much easier.

Nature of the Constituency

The constituency refers to those persons who are impacted by the activities of an organization. They include the customers and clients directly served as well as persons indirectly affected by actions of the organization. We often fail to recognize that the constituency of an organization like Extension is multifaceted. As demonstrated in Chapter 4, the clientele of Extension are quite varied, including such persons as farmers, children, homemakers, and government officials. These multiple constituencies have their own expectations of what the organization ought to be doing. Each special interest group then judges the performance of the organization on the basis of the goals of interest to them and decides whether to support the organization on that basis.

Thompson (1967) has suggested that the way in which these different and possibly conflicting interests are resolved is through the formation of a "dominant coalition." This coalition of constituencies sets priorities and serves as a spokesperson for the organization in the acquisition of resources. The success of the organization in programming and resource acquisition is dependent upon the ability of special interest groups to maintain a workable alliance. Coalitions function as long as individual interests do not overshadow the need to serve the varied interests represented. The assumption is that special interests benefit by uniting in their effort. The make up of the coalition changes over time depending upon which interest groups are represented and the relative influence of each.

Extension's constituencies vary from the elite to the powerless. The relative influence of different constituencies is affected by their relationship to and influence on the process of resource allocation. Constituencies held in high esteem are more likely to have access to and to be heard by policymakers. For example, owners of agribusinesses and large farm operations have traditionally exerted considerable influence on the legislative process at the national level. In contrast, proponents of programs that serve the needs of low-income clients have generally been less successful at affecting support for their cause. Over the years, organizations that have served a more elite constituency have tended to receive better support than those directed at individuals of lower socioeconomic levels (Price, 1968).

In addition, the degree to which the constituency is organized affects its influence. Organized expressions are generally more powerful and successful than large numbers of unorchestrated individuals.

Experience Base

Clients' encounters with an agency are expected to affect their willingness to see the organization supported. People develop a general impression about the organization that is a composite of their specific experiences and the common stock of public knowledge. We reported in Chapter 5 that Extension enjoys a highly favorable rating by clientele.

Though an agency may be doing a good job of performing its intended functions, it has been found that satisfaction with the service does not necessarily translate directly into support for the agency (Warner et al., 1975). Katz and associates (1977) find that people's specific experiences with a service are more positive than their general attitudes about the agency. More specifically, the general feeling toward the agency is negative if individuals have a specific experience with the agency that is negative, but a positive experience does not necessarily translate into a positive attitude toward the agency. In other words, "A negative experience with an agency lowers one's general evaluation of government, but a positive experience does not raise it" (Katz et al., 1977:186). Persons with positive experiences and persons with no experiences with an agency are found to give essentially the same expression of general support, while those with negative experiences give lower general readings. Therefore, specific experiences tend to transfer into a person's general assessment when those experiences are negative.

Historical Support Base

America's farm and rural population has traditionally been seen as the backbone of Extension's base of support; however, shifts in the population have diminished rural and farm political influence. At the turn of the century, rural and farm people comprised a majority of the population, but now only one in four citizens live in rural areas, and only three percent live on farms. These changes have resulted in fewer rural voters with accompanying declines in predominantly rural congressional districts as well as fewer influential congressmen and senators from rural districts serving in leadership roles on key committees. In the decade from 1966 to 1976, the number of predominantly rural congressional districts declined by 95 or to less than 20 percent of the total (Table 6.1).

TABLE 6.1
Trend in Characteristics of House Districts, 1966-1976

Characteristic[a]	1966	1968	1973	1976	Change, 1966-1976
Urban	106	110	102	137	+31
Suburban	92	104	131	114	+22
Rural	181	155	130	86	-95
Mixed	56	66	72	92	+36

[a]The criteria for categorization of the districts are: 50 percent or more of population in standard metropolitan statistical area (SMSA) central city ("urban"); 50 percent or more of population outside central city but within SMSA ("suburban"); 50 percent or more of population outside SMSA ("rural"); less than 50 percent of population in any of three above categories ("mixed").

SOURCE: Congressional Quarterly Weekly Report, 1974:878 and Peters, 1978:28.

There has also been a fragmentation of the farm bloc of pre-World War II days. The political organization of agriculture has moved from reasonably consistent legislation, with broad social purposes generally supported by the society, to narrow special interest legislation (Bonnen, 1965). There has been a movement away from broadly based policy issues toward narrowly defined commodity-specific legislation.

Support for Extension in the U.S.

From the above discussion on factors related to support of a public agency, Extension would be expected to fare pretty well. However, how extensive should support be expected to be? If nearly 90 percent of the population have heard of Extension programs and yet only 27 percent have used the service, what portion of users and the general public are willing to see Extension supported with public monies?

In the national study, individuals who indicated that they themselves or members of their families had used the services of Extension were asked to indicate their support for the agency. They were asked whether there ought to be less, more, or the same level of government support for Extension. For the nation as a whole, four out of five clientele would like to see support at least as great as it is presently, with 39 percent wanting an increase in spending on Extension (Figure 6.1). Only 18 percent of users desire a reduction in support.

FIGURE 6.1
Support for Extension by Users of the Service

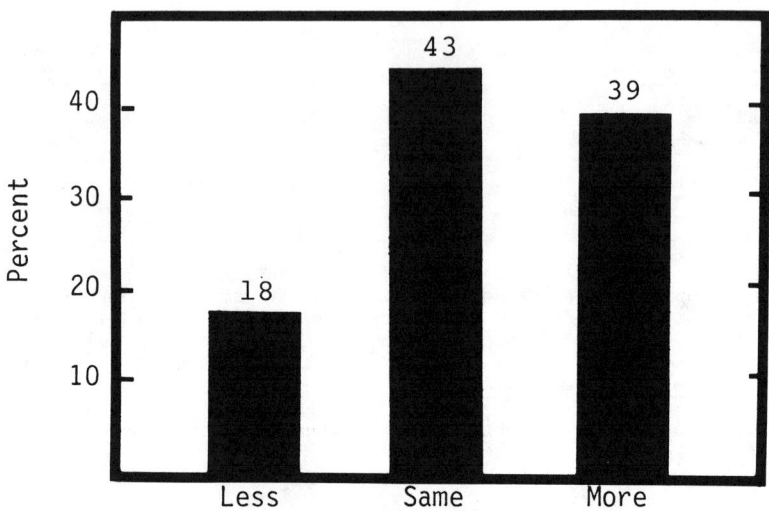

The support for the four program areas—agriculture, home economics, 4-H, and community development—follows a fairly uniform pattern, although users want somewhat more resources spent on agricultural programs than they do on home economics (Figure 6.2). Nevertheless, not more than 21 percent want reduced spending on any of the four programs.

Support for Extension is uniformly high for all regions of the U.S. (Figure 6.3) with slightly greater support registered in the South (90 percent want the same or more spent) and lower in the West (73 percent). However, differences in regional support for the program areas surface when one examines those who want reduced spending (Table 6.2). For example, 21 percent of users from the Western region want support for agricultural programs reduced, compared with only 4 percent wanting a lower level in the South. This same regional pattern is evident for the community development programs.

No differences were found in the desired support for Extension by different types of individuals. The young and old, male and female, rich and poor, black and white, the educated and uneducated, and the rural and city resident were all equally supportive (Appendix B, Table B.6). Likewise, support was uniformly high among all types of users for the four program areas. One might expect certain programs to have greater support among certain segments of the population, i.e., the

FIGURE 6.2
Support for Extension Programs

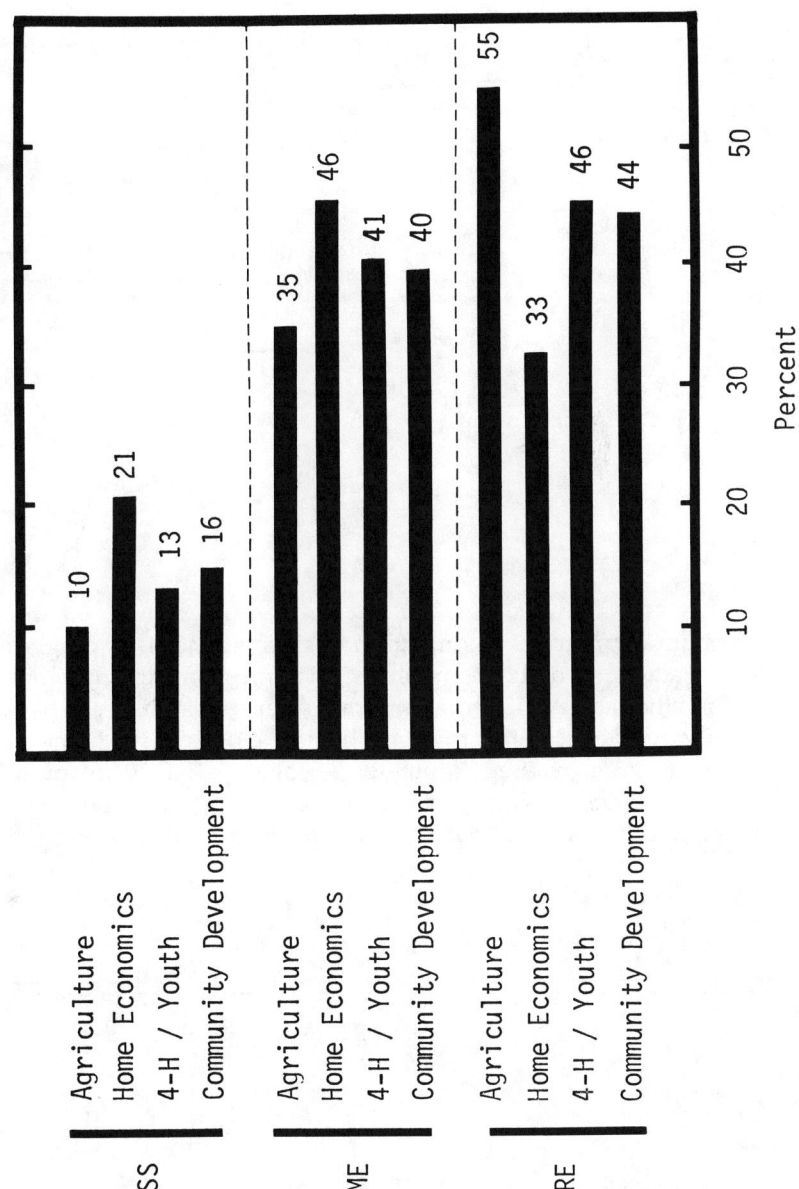

FIGURE 6.3
User Support by Region of the U.S.

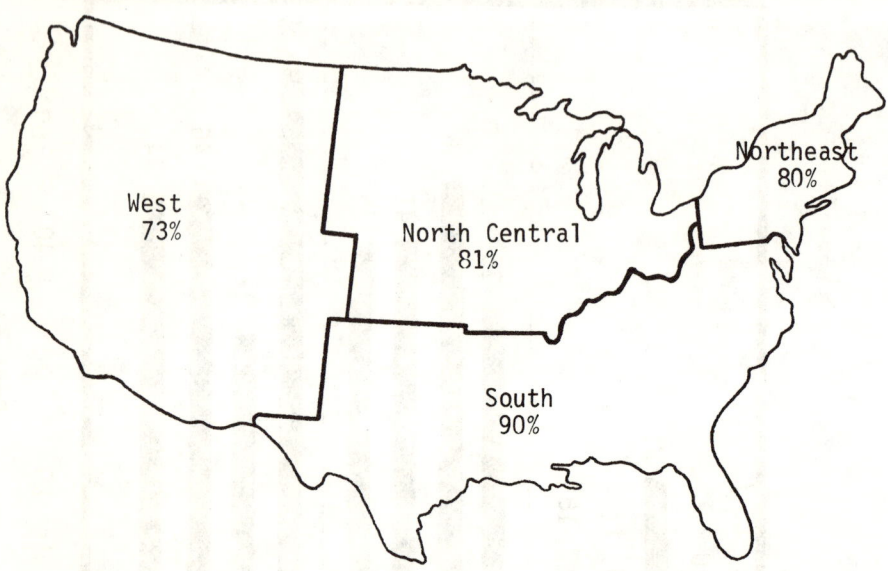

agricultural program among farm residents, the 4-H program among young adults, and the home economics program among women. Such was not the case. Agricultural programs were just as likely to be supported by city residents, 4-H programs by the elderly, and home economics programs by men. (See Appendix B, Tables B.7–B.10 for more details.) Therefore, one can conclude that there is support among all segments of the population and that there is no more support among traditional

TABLE 6.2
Users Wanting Less Spent by Region of U.S.

	Northeast %	Midwest %	South %	West %	χ^2
Extension Overall	21	20	11	27	0.84
Agriculture	15	7	4	21	12.53*
Home Economics	24	18	18	32	4.84
4-H	11	10	14	20	3.95
Community Development	25	11	9	29	13.85*

*$p \leq .05$.

TABLE 6.3
User Support by Level of Satisfaction with Extension in the U.S. (N = 235)

Support	Level of Satisfaction		
	Disatisfied	Satisfied	Very Satisfied
	%	%	%
Less	15	18	11
Same	69	46	32
More	16	36	57

x^2 = 11.9; Gamma = .34.

farms and rural audiences than there is among nonfarm and urban residents.

We generally examine the findings in search of significant differences, when the lack of differences may be what is important. For example, farmers would be expected to be more supportive of Extension than would residents of urban areas, but they are not. However, the fact that farmers were found to be less supportive of other educational programs (primary and secondary schools and libraries) than were nonfarmers suggests that their lack of Extension support may be due to a lower value placed on education or merely a more frugal attitude toward public spending in general (Coughenour et al., 1980). Second, it is often assumed that persons of higher socioeconomic levels (as indicated by income and education) are more supportive of public services like Extension than are persons of low income and educational levels. People do not become more generous in their support as they become better off, not for Extension nor for other community services. Third, though not statistically different, nonwhites are consistently more favorable toward increased spending on Extension than are whites. However, nonwhites generally want more spent on other community services as well.

How Satisfaction Affects Support

How does satisfaction with Extension translate into support? Overall, there is a modest relationship between satisfaction and support (r = .18). As demonstrated in Table 6.3, people are more willing to see support go to Extension if they are satisfied with its performance. Though there are very few dissatisfied users in the national study (5 percent), it is clear that as satisfaction increases so does support (gamma = .34).

TABLE 6.4
User and Nonuser Support for Kentucky Extension by Level of Satisfaction
(N = 9,447)

Support	Level of Satisfaction		
	Dissatisfied	Unsure	Satisfied
	%	%	%
Less	37	21	11
Same	47	70	73
More	16	9	16

x^2 = 511.0; Gamma = .29.

The proportion of people wanting increased support for Extension ranges from a low of 16 percent of dissatisfied persons to a high of 57 percent of the very satisfied.

This finding does not support Katz's observation that a person's positive experience with a public agency is not reflected in his/her expression of agency support. The conclusion here is that positive experiences can result in increased support.

Support: A State Perspective

The Kentucky study relates satisfaction and support for both users and nonusers. The state data reflect a relationship between satisfaction and support similar to the national study (gamma = .29), but one more in line with the Katz finding. Those persons who are satisfied with Kentucky Extension are no more likely to want *more* funds allocated to the agency than those who are dissatisfied (Table 6.4). However, on the other extreme, dissatisfied persons are more likely to want support cut. While only 11 percent of those satisfied want less money spent, 37 percent of those who are dissatisfied want government support of Extension reduced.

Most users and nonusers (about 70 percent) want support for the Kentucky Cooperative Extension Service to remain about the same; they neither want substantial increases nor cuts in support (Table 6.5). However, differences do exist between users and nonusers for those who want either less or more spent. Proportionally, twice as many users want more money allocated when compared with nonusers, while twice as many nonusers want funding decreased when compared to users.

TABLE 6.5
Support for Kentucky Extension among Users and Nonusers (N = 9,854)

Support	Use	
	Users	Nonusers
	%	%
Less	11	22
Same	70	68
More	19	10

x^2 = 259.1; Gamma = .34.

Once people use the services of Extension, they are more favorable toward increased spending and opposed to spending reductions; but the large majority wants the funding level to remain unchanged. Therefore, the relationship between use and support is not particularly strong (r = .23).

The more frequently people use the services of Extension, the more willing they are to see it supported. This is the same type of relationship found between frequency of use and satisfaction. Twice as many frequent users, compared with occasional users, want more money spent on Extension (Table 6.6). Ninety-three percent of those persons who make frequent use of Extension want support to be at least as much as it is now (only 7 percent want it reduced). Thus, there is support among occasional users, but it is not as great as it is for frequent users.

As in the national study, there are few characteristics of either users or nonusers of Kentucky Extension that are related to their support of the organization. For example, farmers are no more supportive than

TABLE 6.6
Support by Frequency of Use of Kentucky Extension (N = 9,854)

Support	Frequency of Use		
	None	Occasionally	Frequently
	%	%	%
Less	22	17	7
Same	68	71	69
More	10	12	24

x^2 = 508.3; Gamma = .34.

nonfarmers, and one is as likely to get support from urban residents as people in rural areas. If there are individuals that express more support than others, they are low-income persons, nonwhites, and those with a low level of education. However, this low socioeconomic group favors increased spending not only for Extension but for most all public services. They also are some of the least powerful in influencing the allocation of resources.

The results of the study of support for Extension demonstrate some findings that are very consistent and others that are mixed. Clearly, users of Extension are more supportive than nonusers, the level of satisfaction with the organization is related to the level of desired support for it, and personal characteristics of users and nonusers are not related to support of Extension. What is unclear is how satisfaction is related to support.

Issues of Support for Extension

Users Are Supportive

Is there support for Extension among its users? The answer is yes. Eight out of ten clientele want Extension's support to be at least as great as it is now, and four of those eight want it increased. Only two in ten want support cut. This level of support is uniformly high in all regions of the U.S.

People find the services of Extension sufficiently useful that they are willing to see their tax dollars used to support the agency. In an era when the public mood is generally against increased government spending for almost any purpose, the level of support for Extension may, in fact, be high relative to other agencies. The extensive clientele involvement in program planning, the autonomy of county offices, and the compatibility of programs with local interests are seen as contributing to support for Extension.

The General Public Is Supportive

Is there support for Extension from the three-fourths of the population who are not Extension users? The vast majority of the public is willing to see Extension's support remain unchanged. One person in ten wants support increased and two in ten want it reduced. Therefore, there is general sentiment for at least maintaining the present level of Extension support. However, there is greater support (and fewer that want support reduced) among users than nonusers. Support increases when persons become users.

In addition, the more frequently people use Extension, the more supportive they are. When it comes to supporting the agency, occasional users respond more like nonusers than they do frequent users. Though there is support to be found among occasional users, the strong base of support lies with individuals who make repeated use of the service.

Traditional Users Are Not More Supportive

Farmers and rural residents are *not* more supportive of Extension than are nontraditional audiences. One is as likely to receive expressions of support from nonfarmers as farmers and from people in urban areas as people in rural areas. Historically, it has generally been concluded that Extension's primary support base originates from farm and rural residents; however, support in urban areas may be stronger than previously recognized. Nevertheless, even with this expression of public support, it is unclear how this public sentiment relates to organized efforts to influence funding sources.

People Support What They Find Useful

How is satisfaction with Extension related to support? Overall, satisfaction is positively related to support but maybe not as strongly as one might expect. The national study found that satisfied people are more supportive, but the state study showed only that satisfied persons are less likely to want funds reduced. Whether satisfied users want more money spent on Extension, or whether they just do not want funds cut, is unclear. Nevertheless, in either case, one can conclude that positive experiences with Extension result in positive opinions about support.

7
The County as the Focus of Service Delivery

The basic unit of the Cooperative Extension Service is the county, for it is here that programs are made and Extension teaching is done.
—Loomis, 1953:58

Most people would agree that the county is truly the focal point of the Extension program. Though national- and state-level issues affect the nature of local programs, the county is where the program all comes together. The federal and state structures are present to support the program that is delivered to people at the county level. Sometimes state and federal officials forget this fact and attempt to treat the program as if it were a single uniform national or state program being delivered to a homogeneous audience. That conclusion couldn't be further from the truth. The counties across the nation are populated with individuals with unique needs and circumstances. They are a diverse people that identify with unique places in which the environmental, demographic, economic, social, and cultural conditions vary substantially. Programs that Extension develops have to reflect this variety of local needs.

In previous chapters, we have been concerned with the individual citizen's assessment of the Cooperative Extension Service. We have been primarily interested in whether the characteristics of individuals are related to their assessment of the service. In doing so, we have ignored the differences that are inherent in the county-based program. In this chapter, we switch the focus from the individual level to the county. We now apply the same Systems Effectiveness Model to county programs.

County-level information is generally available on indicators of inputs and program activities from EMIS and other internal reporting systems; however, data are not readily available on the public's assessment of county programs. The collection of such information for all of the counties in the U.S. would be a very extensive and expensive undertaking.

Nevertheless, an evaluation of Extension would not be complete without an examination of county programming.

In an effort to minimize the variability of different state guidelines, budgetary constraints, and programming resources and, at the same time, to conduct a study of manageable proportions, the county-level study is restricted to the 120 counties of Kentucky. Though the counties of a single state cannot be assumed to be representative of the many geographical, cultural, and economic conditions found in all of the counties of the United States, sufficient diversity does exist in order to demonstrate the important relationships of the study. The Kentucky counties provide a variety of rural and urban locales, distinct socioeconomic areas, and a balance between agriculture and industry in one of the larger Cooperative Extension Services in the United States.

To obtain needed public input for each of the Kentucky counties, 11,015 individuals were surveyed with about 90 respondents reporting from each county (Christenson and Warner, 1982). This in-depth survey of each county provides the basis for the assessment of county programs.

The Historical Context of County Programs

Field staff existed before there were state specialists and federal staff. From the very start, Extension emphasized service to local residents. Efforts concentrated on demonstrations on producers' farms and the formation of 4-H and home demonstration clubs in members' homes. Fundamental to the Extension concept has been the placement of professional staff in close proximity to the people to be served. Even today, over 70 percent of Extension staff are officed at the county level.

The predominate model of county staffing is to locate at least one professional within the county boundaries. In most states, there are several county staff working in each county along with supporting secretarial and/or administrative staff. Usually, the staff is comprised of an agricultural agent, a home economics agent, and a 4-H agent. In recent years, a community development agent has been added to a few counties. Often, though, community development programs have been handled by other county or area staff. In more populous counties, a county's Extension staff may number as many as twenty or more individuals, but the average number of professional staff within counties across the United States is 3.7 full-time equivalent positions.

Though the relationships vary from state to state, county Extension offices function somewhat independently of the state and federal offices. Clearly, the county office does not function as a franchise or a branch office of a large centralized organization. The county office is established to respond to local needs while drawing upon the support and expertise of the state and federal governments. Since county needs across the United States vary dramatically, Extension programs also differ significantly. However, the same general philosophy and methods prevail throughout most state and county units. As DeMarco (1980:4) observes, Extension programs from county to county often vary "in tone rather than in substance."

The ecological, demographic, and social conditions of the 3,150 U.S. counties vary considerably. In the midwestern and southern portions of the United States, counties tend to be small and the population is fairly densely populated. Whereas, in the northwestern and southwestern part of the United States, many counties are more than 6,000 square miles and are sparsely populated. Thus, it is difficult to talk about counties throughout the United States as similar geographical dimensions and like populations.

In short, considerable variation exists in the functioning of county Extension staff within individual counties throughout the United States. Likewise, considerable diversity exists in the ecological, demographic, and socioeconomic characteristics of the residents of these counties. Since, for Extension, the county is the key unit in the provision of services, one also must understand the county context in which the service is developed and provided.

A County-Based Effectiveness Model

Within the county, the Extension staff utilizes inputs to conduct program activities that result in services delivered. These are the same components as described in the comprehensive Systems Effectiveness Model as discussed in earlier chapters. This chapter will examine aspects from the overall model that relate to county operations.

Four major components comprise this County-Based Effectiveness Model: county inputs, county program operations, public impact, and the county environment (see Figure 7.1). While including four elements of the larger model, this county-based effort has limitations due to the small number and scope of variables selected and the blending of four data sources (Census data, survey data, budget figures, and EMIS information), each of which was gathered for slightly different purposes. However, such limitations should not overshadow the illustrative importance of both the model and findings. No other evaluation effort has been successful in combining county-level information from such diverse sources. Thus, this effort should be seen as an initial attempt open to refinement, elaboration, and modification in subsequent studies. It is hoped that the model will provide a framework for inclusion of other indicators in future efforts.

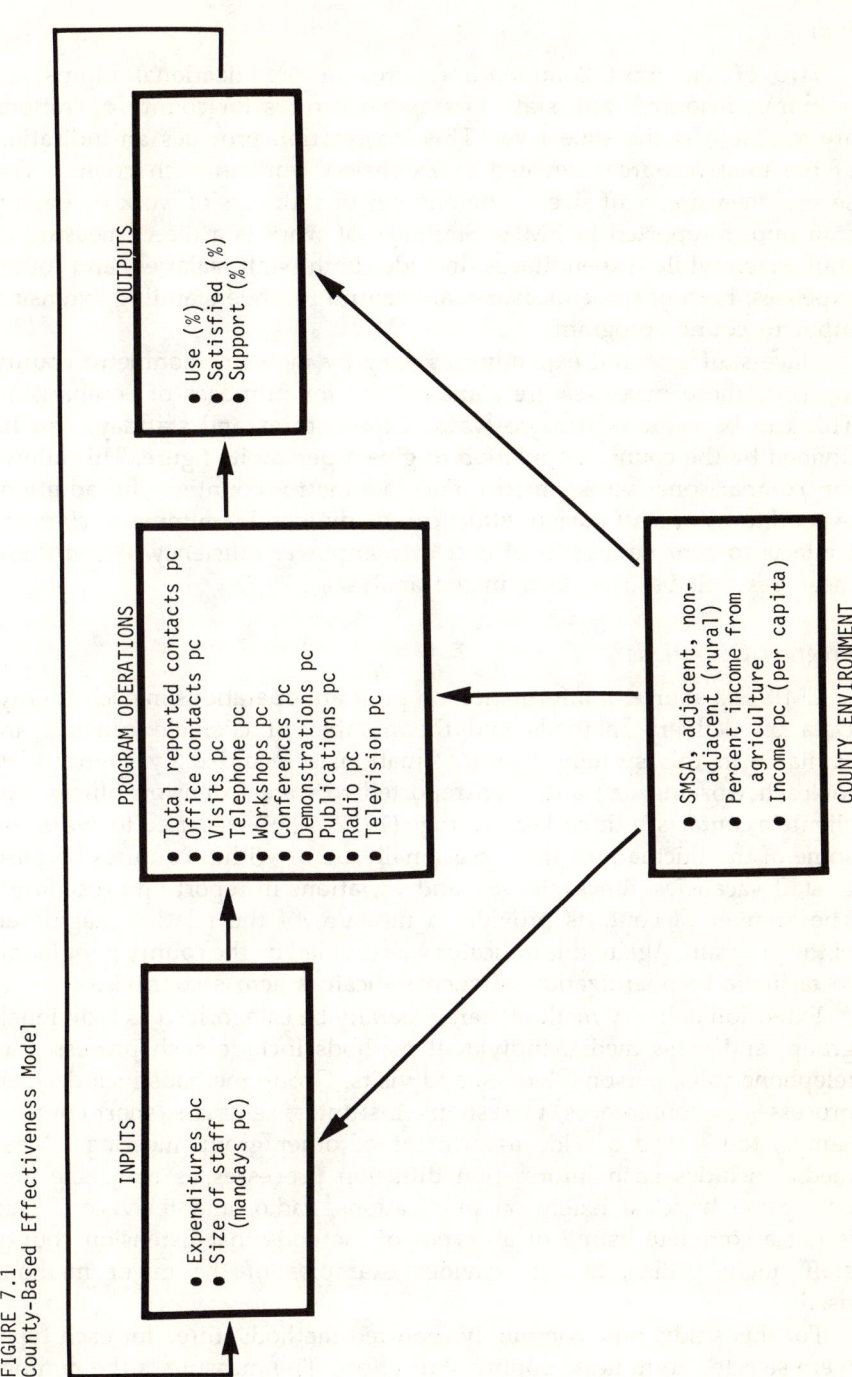

FIGURE 7.1
County-Based Effectiveness Model

Inputs

Two of the most common measures of organizational inputs are economic resources and staff. First, expenditures for county operations are available at the state level. This information provides an indication of the total resources devoted to Extension work in each county. The second measure, staff size, is the number of staffdays of work by county staff and is reported in EMIS. Staffdays of work is a direct measure of staff size, while expenditures include both staff salaries and other expenses. Both of these measures are central to understanding Extension input to county programs.

Since staff size and expenditures vary by metro or nonmetro county location, these measures are standardized for purposes of comparison. This can be achieved in two ways. Expenditures and staffdays can be divided by the county population to give a per capita figure. This allows for comparisons across metro and nonmetro counties. In addition, expenditures or staff days of effort can be divided by number of clientele contacts to achieve a ratio of cost or manpower efficiency. All of these measures will be used later in the analysis.

Program Operations

EMIS is a source of information on program operations in each county. Data on delivery methods and the number of clientele contacts are available in this system. This information is recorded by county staff for each working day and then reported to state Extension offices. For clientele contacts, a three-year average (1977–1979) was used to overcome some of the fluctuations that occasionally occur within counties because of staff vacancies, illness, leaves, and variations in reporting procedures. The number of contacts provides a measure of the relative magnitude of the program. Again, this indicator was divided by the county population to facilitate standardization of such indicators across counties.

Extension delivery methods can generally be categorized as individual, group, and mass media. Individual methods include such processes as telephone calls, personal letters, and visits. Group methods include such processes as conferences, workshops, institutes, seminars, short courses, camps, tours, and a wide assortment of other group meetings. Mass media includes such information diffusion processes as magazine and newspaper articles, Extension publications, radio, and television. This is not a complete listing of all types of methods that Extension county staff might utilize, but it provides examples of the major methods used.

For this study, nine commonly reported methods, three for each type, were selected to indicate county staff effort. The measure is the number of hours devoted to each method and was calculated as the yearly average for the three-year period, 1977–1979, then divided by the county population to provide a per capita indicator. The three individual methods selected were *office* contact (a visit to a county Extension office by a client to obtain or give information), *telephone* calls (a telephone contact made by a client to an Extension office or staff member, or by staff to a client to obtain or give information), and *visits* (a personal call on a client by a county staff member to give or obtain information).

Group methods include conferences, workshops, and demonstrations. *Conferences* denote a general type of group meeting held in a single location for the purpose of hearing prepared presentations and discussions about one or more educational topics. *Workshops* indicate a special interest activity of one or more meetings in which each participant actively studies or works on a project of particular interest to him/her in the subject matter being taught by qualified instructors. *Demonstration* is a method used to verbally or visually explain various practices or processes.

Mass media methods include radio, television, and publications. *Radio* reflects the staff time devoted to developing and/or giving a radio message on subject matter or Extension activities. *Television* is a similar measure except that it involves the use of film or videotape. *Publications* involve the number of hours devoted to the development of written material for presentation of information of some subject matter or on Extension itself.

Public Assessment of Output

Effort is not the same as effect. In addition to delivering a program, the effort must meet the needs of a target audience. Three indicators of the client's reaction are contained in the model. They are use, satisfaction, and support, and all three are taken from county survey results.

Use of Extension's services provides an assessment of how extensively the public within a county avails themselves of the service. The question read: "In the past year have you or your family *used* or contacted the county Extension agent?"[1] The variable for this county-level study was coded as to the percent of respondents within the county who said that they or a family member used or contacted Extension during the previous year. It should be recalled that this is a use measure for *only* the previous year. As noted in Chapter 4, this figure generally could be seen as a conservative measure of public contact with Extension county staff.

The next measure for public assessment of county Extension output focused on the percent of respondents in the county who said that they were satisfied with the Cooperative Extension Service. This percentage does not include those who were dissatisfied or undecided (unsure).

Finally, to obtain an indicator of public support for Extension within the county, the percentage of respondents in each county who indicated they felt that the same or more government funds should be spent on the Cooperative Extension Service was calculated.

For each of these measures of use, satisfaction, and support, any respondent who indicated they did not know what Extension represented was excluded from the analysis. Approximately 10 percent were excluded for this reason. This is quite close to the national level of 13 percent who reported being unaware of Extension and/or its programs (Chapter 3).

Environmental Factors

Three variables were selected to indicate the demographic and socioeconomic conditions of the 120 Kentucky counties. These variables were selected because they are readily available from the Bureau of the Census and the Census of Agriculture. Thus, such information can be easily obtained in other states and counties. In addition, these variables are of central concern to Extension's original mandate and the focus on congressional directives for service to special groups. They are also consistent with research findings for other public services (Bonjean et al., 1969; Christenson, 1976; Christenson and Taylor, 1982; Lyons, 1977).

A Standard Metropolitan Statistical Area is a county designation related to the county's inclusion in a metropolitan area. A threefold classification scheme is used in this study: (1) a metropolitan county, (2) a nonmetropolitan county but adjacent to a metropolitan area, and (3) a nonmetropolitan nonadjacent county. This three-level classification scheme was adopted because it appears to represent best the reality of service delivery areas for the counties.

Per capita income provides an indicator of economic well-being of county citizens and is readily available from secondary information sources. Percent income from agriculture suggests the degree to which agriculture contributes to the economy of a county and suggests the relative degree of influence of agricultural interests in local decision making.

Demonstrating the Model with Kentucky Data

In this chapter, the county Extension program is the focus of attention; therefore, all of the indicators will be aggregated for each of the 120 counties. In Kentucky, each county is fairly small in land area. The average size is 330 square miles and ranges in population from 3,200 to over 600,000. There are 17 metro (SMSA) counties, 34 nonmetro counties that are adjacent to metro centers, and 69 nonmetro nonadjacent counties. Kentucky has a strong agricultural base, and, while one might expect the nonmetro counties to have a higher proportion of income from agriculture than the metro counties, such is not the case. Many of the metro areas serve as marketing outlets and supply centers for agriculture, thus reflecting agriculturally related income. In short, no relationship is apparent in Kentucky between SMSA location and percent of county income from agriculture. (Correlation coefficients for county-level measures can be found in Appendix B, Table B.11.)

Per capita income levels of county residents are as one might expect: higher in metro counties than in nonmetro areas. For example, if 120 counties are divided into quartiles according to per capita income levels, 81 percent of the metro counties fall into the highest income bracket, while only 17 percent of the nonmetro counties fall into this category (data not presented). There exists only a slight relationship between the income level of a county and the percent of income from agriculture.

Resources

Because of its educational focus, county Extension work is labor intensive. Most of a county's Extension budget is for staff salaries.[2] As a result, the size of county staff has a very strong correlation with county expenditures ($r = .90$). Essentially, the two variables can be equated in this study. Because metro counties have large populations, the expenditures per capita for these counties is lower than for nonmetro counties. Low income and high agricultural income counties have higher Extension expenditures per capita.

Program Implementation

The overall distribution of staff time by type of educational method shows 54 percent of the time being used for individual contacts, 40 percent for group methods, and only 6 percent for mass media. It is also interesting to note that personal visits to clientele by county staff represent 39 percent of the total. The average distribution of time among all methods for county staff in Kentucky for three years (1977–1979) is reported in Table 7.1.

Relating EMIS Contacts to Survey Estimates

How do agency reports of contacts relate to general population surveys concerning use of Extension? The relationship is moderate in strength ($r = .53$). It should be recalled that agency-reported contacts (EMIS) count people each time they make contact with a staff member, while survey data report the number of clients (persons). The survey information makes no distinction between single and multiple contacts. Thus, the two variables count clientele in different ways. For these reasons, one

TABLE 7.1
Distribution of Staff Time to Different Educational Methods, Kentucky Counties, 1977-1979.

Individual Methods	Percent
*Visits	38.6
*Office calls	5.0
*Telephone calls	5.0
Individual studies	2.0
Circular letters of newsletters	1.7
Personal letters	1.1
Employee counseling	0.3
Formal interviews or conferences	0.3
Inservice training	0.2
Subtotal	54.2
Group Methods	
Inservice training	9.2
General meetings	7.6
Committee meetings	4.1
Staff meetings	3.6
Camps	2.4
Achievement and recognition events	2.3
*Conferences	2.2
*Method demonstrations	2.0
*Workshops	1.9
Leader training	1.7
Result demonstrations	0.6
Field days	0.6
Tours	0.5
Schools or institutes	0.5
Short courses	0.4
Seminars	0.4
Applied research or field trails	0.2
Telelectures	0.0
Subtotal	40.2
Mass Media Methods	
Exhibits or displays	1.9
*Radio	1.3
Newspaper articles or releases	1.3
*Publications	0.9
*Television	0.2
Magazine or journal articles	0.0
Subtotal	5.6

*Used in subsequent analysis.

The County as the Focus of Service Delivery 109

would not expect the same results, but the correlation of 0.53 does seem to reflect moderate to strong congruence between the two measures. In short, the number of contacts recorded in the EMIS system agrees fairly well with the survey results.

Contacts and Methods

The number of contacts per capita varies considerably in different counties. A first question raised was the extent to which the commonly reported methods such as conferences, workshops, publications, newsletters, radio, and television were related to EMIS reported contacts. The number of contacts was found to be closely related to the use of individual and group methods. This relationship occurs irrespective of the type of county. Inasmuch as the reporting system does not document the number of people reached through mass media methods, those methods were not related to the number of reported contacts (Appendix B, Table B.12).

A similar pattern relating individual and group methods with reported use was found in the population survey. For the most part, the same individual (office contacts and visits) and group (workshops) methods are related to public use patterns. In addition, television is an important new consideration to predict public use. The hours devoted to development of Extension television programs is positively related to higher county use patterns. Finally, lower income counties, which are usually the more rural counties, report a higher county use pattern.

Cost Effectiveness in County Programs

Effort, impact, and efficiency tap three distinct aspects of organizational performance: what the agency does, what clientele derive from it, and how cost effectively the process is performed. To this point, we have tried to assess effort and impact. However, increased demand from decision makers for cost effectiveness measures requires that consideration be given to indicators of efficiency.

How much does each Extension contact cost? The cost per contact was calculated for each Kentucky county. The cost of the county program was divided by the total number of contacts reported by the county staff. The cost of specific county programs in Kentucky ranged from a high of approximately $554,000 to a low of $42,000 in 1979. The average (mean) for the 120 counties was $108,000. The number of contacts in 1979 across the 120 counties ranged from a high of 146,000 to a low of 2,000, with a mean of 22,000. This represents only reported contacts by individual and group methods, since mass media contacts are not reported. The cost per contact in 1979 for Kentucky was $4.91. This figure represents only county cost and contacts; it does not include area and state staff contributions.

Educational methods used by counties were then related to the cost per contact for counties (data presented in Appendix B, Table B.12). Visits by county agents outside the office is the most expensive method in terms of contacts. This is not surprising in that one-on-one personal contacts at the location of the client demand considerable time and resources. As noted earlier, it was the most frequently used method (39 percent of all contacts are by personal visits). Television also appears to be an expensive method. However, since data are not collected or reported on mass media contacts, this provides a misleading finding. If estimates of the number of people reached by a television program were included, a quite different cost assessment would likely result. In addition, when television as a method is adjusted for the county situation, television becomes a cost efficient method. This reflects the fact that urban counties tend to devote more time to television methods for which no contacts are reported. Also, television broadcast areas do not stop at county boundaries; therefore, television programs provide benefits (spill-over effects) for many nonmetro or rural counties at little or no cost to them. Thus, in the larger state context, television can be expected to be quite cost efficient.

All group methods are as cost efficient, if not more so, than individual methods, though telephone contacts also tend to be very cost effective. In short, of the methods studied, visits are the most expensive, and conferences are the least expensive. One cautionary note, the amount of variation in cost attributable to the different methods is relatively low. Thus, the estimates and relationships, while illustrative, do not reveal the complex support system, indirect costs, and complexities of assigning costs to particular program operations.

The County Effectiveness Model

By putting all the information that has been discussed into a county-based assessment model, interrelationships can be studied (Figure 7.2 and Appendix B, Figure B.1). Because of the close relationship between county budgets and staff size, only county expenditures were used. Furthermore, because of the close relationship between methods (hours of staff effort) and number of contacts, only per capita contacts were included.

There is strong relationship between agriculturally based counties and county budgets (expenditures per capita). In other words, in counties where farming is important, there is a tendency to spend more per

FIGURE 7.2
Findings for County-Based Effectiveness Model

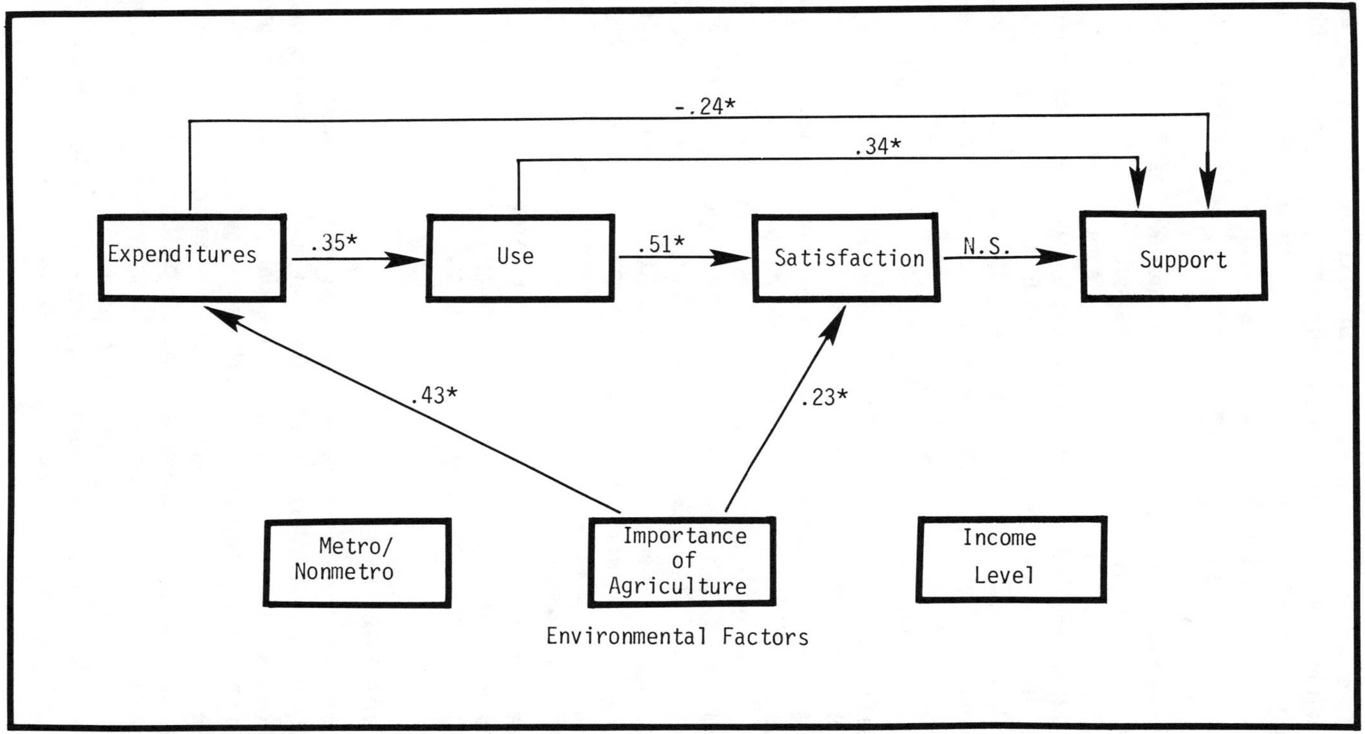

*Standardized correlation coefficients significant at the .05 level.
N.S. = Not Significant.

person on Extension. However, since these counties often have a smaller population base, they do not necessarily have the largest overall budgets. County expenditures are very strongly related to number of contacts and are the most important factor in explaining the percent of county residents reporting use of Extension services. In short, the more money spent, the more people served.

Use of Extension was greater in agriculturally based counties (percent income from agriculture) and counties with lower per capita income. These two measures (income per capita and percent income from agriculture) are indirect references to rural counties that characterize these situations. In short, percent of public use of Extension services is most influenced by county budgets and the rural county situation.

Satisfaction with Extension programs, according to county residents, was most influenced by use of Extension services and, to a lesser extent, by counties with an agricultural base. In other words, county residents were more likely to be satisfied with Extension if they used the programs and if they lived in agricultural counties.

The percent of county respondents who wanted the same or more governmental funds allocated to Extension (as opposed to those who wanted less funds allocated) also was most influenced by use. A slight negative relationship was noted between support and present level of expenditures. The greater the present level of expenditures, the less people want to spend more. This suggests that urban counties that have had lower expenditures per capita are more supportive of more spending than are rural counties where expenditures per capita are higher. However, overall, the model provided only limited information as to what contributes to developing public support of Extension. The relationships examined do a pretty good job of explaining expenditures, use, and satisfaction but don't adequately predict public support.

County Summary

Individual contacts, especially personal visits, account for half of the total time allocation by county Extension staff. As a result, individual methods such as office contacts, personal visits, and telephone calls combined with group methods such as conferences and workshops are important variables for predicting both the reported number of clientele contacts from agency records and the use patterns as reported in general population surveys. Interestingly, the county situation, which includes metro-nonmetro location, level of income, and the relative importance of agriculture, seems to have little or no effect on educational methods used. This finding confirms DeMarco's observation that county programs differ more in tone than in substance.

The number of agency-recorded contacts is moderately to strongly related to respondent-reported use patterns, thus lending increased confidence to agency reports. Both the number of contacts recorded by Extension and the percent of the public who consider themselves Extension users are primarily influenced by county budgets. The greater the expenditures per capita, the greater the number of reported contacts and the higher the pattern of use. Use of the Cooperative Extension Service is primarily influenced by staff size and level of expenditures. This relationship holds for both the agency's indicator of client contacts and the results of the general population survey. As more money is spent, more clientele are reached. However, expenditures are not shown to be related to the level of satisfaction and are negatively related to support. On the other hand, use is strongly related to both satisfaction and support. In short, those persons who have used the services of Extension are very satisfied with the programs and are supportive of the organization.

As one would expect, one-on-one communication methods proved to be the most expensive form of contact with clientele. This was especially true if the contact included a visit to the location of the client. Group methods and mass media efforts offer cheaper alternatives. However, the choice of appropriate educational methods must be made not only in light of cost considerations but also according to the nature of the message and the intended audience. For example, lower educational achievers in such programs as the Expanded Food and Nutrition Education Program may need a higher level of personal contact. Nevertheless, Extension staff need to reassess continually the balance between the different educational methods used. There is the general feeling that the agency relies too heavily on personal methods.

The county-based assessment model illustrated with the Kentucky Cooperative Extension Service provides a framework for integrating county-level information from such diverse sources as census data, budgets, EMIS, and population surveys. It suggests one avenue for investigating the impact of Extension on its environment. Findings of the study can aid decision makers in the allocation of scarce resources at the federal, state, and local levels. It also assists in the monitoring of efficiency of service delivery.

However, many questions remain to be answered. For example, in what way does public support and/or use patterns influence funding sources? How does public sentiment and use influence various lobbying groups? The model proposed and the questions raised at best set parameters for future investigations. More extensive policy and programming implications of the county-based assessment model will be discussed in the concluding chapter of the book.

Notes

1. The term Extension agent in Kentucky is a title that refers to all county staff.

2. Except for inflationary increases, expenditure data (e.g., county budgets) are very similar from year to year. A comparison of expenditures for the years 1976 to 1980 for each of the 120 Kentucky counties showed zero order correlations of over .97. Essentially, monies are allocated irrespective of open positions and other fluctuations. Actual staffdays of effort more closely reflect year-to-year fluctuations in county situations. The zero order correlation between average staff effort over a three-year period and expenditures is .91. However, staff effort comparisons between years drop below 0.80.

8
Summary and Priorities

The Cooperative Extension Service has an annual budget in excess of $800 million. It has a staff of close to 17,000 persons, three-fourths of whom are located in the more than 3,000 counties across the United States. The size of the organization has remained fairly constant for the last 25 years, with a growth rate of staff between 1958 and 1978 of less than one percent per year. Approximately one-third of staff effort is devoted to agriculture, one third to home economics, and one-fourth to 4-H, and one-ninth to community development. In short, Extension is one of the largest educational outreach organizations in the United States. It has a budget larger than many federal agencies and staff who are located in most counties throughout the United States. With an organization of this magnitude, it is no wonder that attention is being focused on its accountability.

In this chapter, we attempt to summarize what we have learned both about the organization and the impact of the organization upon its environment. The Systems Effectiveness Model provides a framework for systematically viewing an organization, while at the same time generating specific findings about the perceived effectiveness of the organization. Previous evaluations of Extension have tended to focus on single programs such as 4-H or activities such as the Expanded Food and Nutrition Education Program (EFNEP). Such evaluations have focused primarily on inputs and program activities. In this study, we have focused on Extension as a total organization. In the following pages, the specific aspects of the model are interrelated and summarized (Figure 8.1), and future program priorities are presented. The purpose is to summarize trends, suggest future directions, and establish a foundation for discussing policy issues in the final chapter.

Awareness, Service, and Support

Eighty-seven percent of the U.S. population are aware of Extension programs. While 4-H programs have the highest level of visibility with

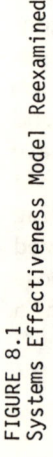

FIGURE 8.1
Systems Effectiveness Model Reexamined

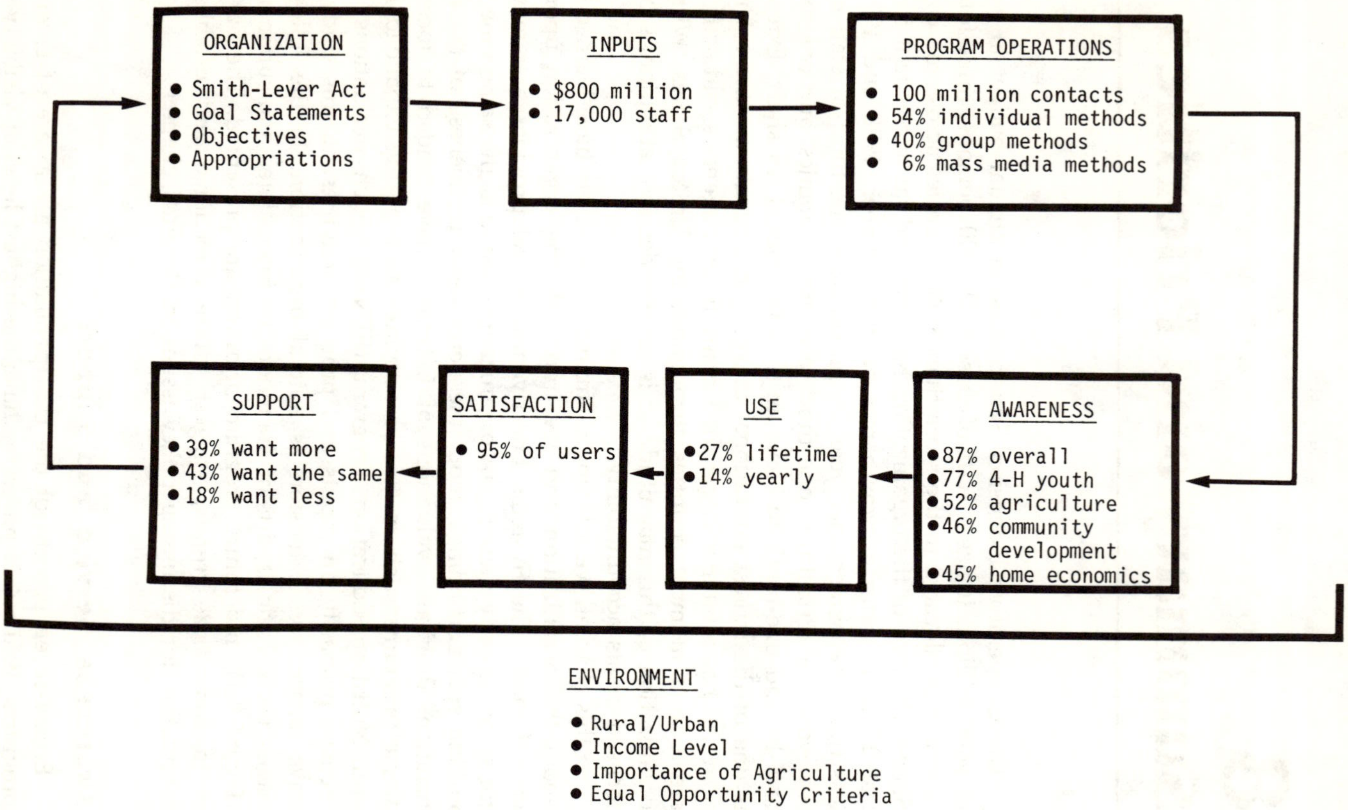

Summary and Priorities 117

the general public, close to half of the population recognize components of the four program areas. However, it is clear that Extension conveys different images to different people. Extension has multiple identities.

Extension touches the lives of a large segment of the American population. According to EMIS information, Extension made over 100 million contacts during 1982. Since this counting system can, and often does, include multiple contacts with the same individual, it is difficult to relate this figure to the total population. According to the national survey, members of 27 percent of the households in the United States have used Extension services sometime in their lives. In 1981, over 11 million households throughout the U.S. were involved in Extension programs. While no differences were found in the degree of awareness of Extension among the various segments of the population or of the different regions of the country, there were clearly some different patterns in use of Extension programs.

Little or no overt discrimination is apparent in Extension program efforts according to race, though the organization is serving a lower proportion of nonwhites. No discrimination was noted in use patterns according to age or sex. While Extension continues to serve proportionally more rural people and more farmers, numerically Extension serves more people in metropolitan America. Extension's image as a rural agency seems to be somewhat inconsistent with the fact that two-thirds of its clientele live in metro areas. Of course, three-fourths of all U.S. citizens live in metropolitan areas.

Extension serves middle-class Americans. Most Americans are middle class, and Extension seems to be directing its programs at the mainstream of the public. On the two extremes, upper income, highly educated, and whites seem to take more advantage of the services provided by Extension, while low-income groups and certain other segments of the public seem to be slightly underrepresented in most programs. However, a notable exception is that low incomes and nonwhites are proportionally overrepresented in the community development program.

How do clientele report using Extension programs? Over 90 percent of the users receive printed material from Extension and/or listen to radio and T.V. programs conducted by Extension staff. Approximately 40 percent of the more than 11 million households who used Extension programs during 1981 attended a meeting or workshop. Most used Extension more than once. That is, if they had one contact with Extension, they were more likely to access the service again. This was particularly true among rural residents and farmers. These two subgroups make more extensive use of Extension.

Users are satisfied with the services they receive. Nine out of ten Extension *users* are satisfied with the services of Extension, and that

TABLE 8.1
Summary of Major Findings

Largest educational outreach program in the world
- $800 million budget and 17,000 staff
- Staff in most all counties in U.S.
- Size of organization relatively stable
- Cooperative funding among federal, state, and county governments

Extension's educational mission
- Conduct a voluntarily accessed educational program
- Serve the people "not attending or resident in" land-grant colleges
- Programs to include agriculture, home economics, rural life, and subjects relating thereto
- Methods to include demonstrations, publications, giving of instruction, and imparting information
- Conduct other programs as specified by Congress (urban 4-H, Integrated Pest Management, EFNEP, etc.)

Extension's changing environment
- The role of information in U.S. society
- Population shift from rural majority to rural minority
- Decline in proportion of farmers in U.S. population

Extensive program contact
- County-based programs with considerable local input
- Staff report over 100 million contacts yearly
- Number of contacts directly relates to size of county budget

Extension's traditional methods
- Half of staff time is devoted to individual client contact
- Most clientele use printed material and mass media methods
- County situation has little impact on educational methods used

Extension's diverse image
- High level of visibility (87 percent of the public)
- Multiple identities
- 4-H program most visible

Americans use Extension
- One-quarter of households have used Extension
- Over 11 million households used Extension in 1981
- Many clients are multiple users

Equal opportunity exists
- Middle-class orientation
- No perception of overt discrimination (sex, race, etc.)
- Serves low proportion of minorities

Extension serves urban and rural people
- Two-thirds of clientele reside in urban areas
- Higher proportion of rural and farm residents are users
- Many urban users have rural roots

Users are highly satisfied
- Satisfaction fairly constant throughout U.S.
- Satisfaction uniformly high across all programs
- Satisfaction with Extension higher than most public agencies

Nonusers have mixed sentiments
- One-half are satisfied
- One-third are "holding their vote"
- Most unsure reside in urban areas
- One out of seven is negative
- Rural nonfarm somewhat negative

Agricultural program
- Largest program—one-third of staff and budget
- More frequent users
- Greatest support for agriculture

Home Economics program
- Almost one-third of staff and budget
- One-quarter of clients are male
- Support for Home Economics lowest of four programs

4-H program
- Highest visibility
- Serves more upper-middle class persons
- Greatest number of contacts of four programs
- Previous 4-H involvement increases present use

Community Development program
- Smallest program (six percent of staff)
- Serves greater proportion of minorities

Public support present
- Most want support to remain unchanged
- Few want funding cut
- Use closely related to support
- Heavy users most supportive

Organizational support fragile
- No unified voice speaking for Extension
- Loss of rural congressional seats
- Dependence upon agriculture
- Fragmented farm bloc

percentage increases as frequency of use increases. *Nonusers* are split in their perception—about a third really do not know what Extension has to offer, about half are satisfied, while one out of seven is dissatisfied.

Most of the users who are satisfied also want more government funds allocated to Extension. The strongest predictor of support for Extension is whether the person has used the service. Extension seems to have a quality product, and those who partake are satisfied and supportive. However, certain segments of the public who do not avail themselves of the service tend to be rather negative about support.

Summary in Brief

Overall study findings are presented in outline form in Table 8.1. This presentation provides more detailed information than the discussion in the first part of this chapter. The outline offers the convenience of locating a concise statement of the major findings in one place in the book. We expect that the reader will seek out a more detailed discussion of the results in appropriate chapters.

Future Program Priorities

Extension is involved in hundreds of different programs and activities that compete for staff time and resources. The present distribution of effort is known, but what are the future priorities? What program areas or subject matter topics ought to receive greater or lesser attention one year or five years from now? Extension seldom abruptly alters major program directions. Rather, change generally comes gradually and in small increments. The motivations for change are often staff vacancies, increases or declines in funds for designated purposes, and changes in the nature of clientele demands. In the last decade, the distribution of professional resources among Extension's four major program areas has remained relatively unchanged. The only significant shifts in overall emphasis are toward agriculture aned away from 4-H youth. During the seventies, resources devoted to agriculture increased by 2.6 percent, and 4-H declined by 4.5 percent. These changes represent gradual trends that started in the mid-seventies.

For planning and reporting purposes, the federal office of Extension has defined 21 major program components. Table 8.2 reports the distribution of professional staff time by each component and shows the changes that have occurred over a four-year period. The greatest amount of time is devoted to Crop and Livestock Production. Increases have been registered for Crop Production, Organizational Development and Maintenance, and Leadership Development. Declines in staff efforts have

TABLE 8.2
Distribution of Professional Staff Time by Program Component, FY78 to FY82

Component	Percent of Total					
	FY78	FY79	FY80	FY81	FY82	Change[a]
Crop Production	19.4	19.9	20.8	21.4	22.0	+
Livestock Production	11.8	12.4	12.4	12.0	12.0	
Organizational Development and Maintenance	9.1	9.4	10.7	10.3	10.6	+
Leadership Development	8.0	8.2	8.8	8.7	9.3	+
Family Life	8.0	7.3	6.6	6.3	5.8	−
Business Management & Economics	5.5	5.2	5.1	5.2	5.0	−
Food and Nutrition	5.0	5.2	5.0	5.2	5.3	
Ecology, Natural Resources & Environment	4.2	4.1	4.4	4.6	4.3	
Expanded Food and Nutrition Education Program	5.3	4.4	4.1	3.9	3.9	−
Housing and Home Environment	4.1	4.2	3.7	3.6	3.4	−
Personal & Family Resource Management	3.2	3.3	3.1	3.4	3.1	
Agricultural Marketing & Farm Supplies	3.0	3.0	2.8	2.7	2.7	
Textiles and Clothing	3.0	2.7	2.5	2.6	2.7	
Leisure and Cultural Education	2.0	2.4	2.5	2.3	2.1	
Community Services and Facilities	1.6	1.6	1.5	1.5	1.3	
Mechanical Science, Technology & Engineering	1.4	1.2	1.3	1.4	1.3	
Economic, Manpower, & Career Development	1.1	1.4	1.2	1.3	1.5	
Comprehensive Community Planning	1.2	1.3	1.0	1.1	1.0	
Safety	1.2	1.2	1.1	1.1	1.2	
Human Health	1.2	1.1	1.0	1.0	0.9	
Government Operations & Finance	0.7	0.7	0.6	0.6	0.7	
	100.0	100.0	100.0	100.0	100.1	

[a]Direction of change is indicated if it is one-half of one percent or greater over the four-year period.

Source: Annual Reports, Extension Service, U.S. Department of Agriculture.

been reported in Family Life, Business Management and Economics, Expanded Food and Nutrition Education Program, and Housing and Home Environment.

When responding to an informal survey of redirection of program efforts (Tanner et al., 1982), state Extension directors indicated a reallocation of program emphases *toward* Farm Business Management and Marketing, Family Resource Management, Crop/Livestock Production, Conservation/Minimum Tillage, and Food and Nutrition and *away from* Family Life, Home Energy Use, and Natural Resources. Obviously, it is possible for different states to be moving in opposite directions, but some general trends seem to be evident. According to state directors, more emphasis is being placed on food production and distribution and farm and family resource management; while some states are placing less emphasis on family life, conservation and natural resources, and energy programs.

The distribution of professional staff time confirms the increased focus on crop production and natural resources as it relates to tillage practices and the decline in effort spent on family life. Staff time allocation does not confirm an increase in farm and family resource management; in fact, Business Management and Economics registered a decline in staff time. But, there was a reduced effort in EFNEP and housing programs. Leadership and organizational development show an increase, but there is the feeling that they have become catch-all categories for time not easily allocated to other subject areas. Therefore, this increase may not reflect a real change in staff function but, rather, is the result of the nature of the reporting system.

In order to project the priority programs of the future, Extension users identified in the national survey were asked to indicate the importance they would place on some thirteen program thrusts. Though these topics do not exactly conform to the 21 major program components, they were chosen to represent items that clientele would understand from the major programs of Agriculture, Home Economics, 4-H/Youth, and Community and Natural Resource Development. Table 8.3 summarizes the results. Ranked first was Energy Conservation, with nine out of ten Extension clientele indicating it as important for future Extension efforts. Next were Food Production, Human Nutrition, Youth Development, and Natural Resources and Environment. Four out of five respondents saw these as important subjects for Extension. Rated sixth and seventh in priority were Food Marketing and Health Care. Consumer Affairs, Economic Development, Family Life and Personal Development, Community Services and Facilities, and Home Gardening received slightly lower ratings. Only Housing, which was rated last, was considered important by less than half of the respondents.

Natural Resources and Environment was the only area that differed significantly in priority rating across the four regions of the U.S. The western part of the U.S. rated this topic considerably lower than the other three regions. There were no important rural-urban differences in the ranking of program priorities.

The high ranking given Energy Conservation by clientele may be surprising to some people. Energy conservation is an area that has received increased attention since the mid-seventies due to a national energy crisis and the availability of money specifically for conservation programs, but it still represents a very small portion of total staff time. Being a relatively new program thrust, energy conservation is not identified as one of the 21 program components by agency personnel.

TABLE 8.3
Priority for Extension Programs by Users

Program	Percent Indicating Topic Is of Great Importance
Energy Conservation	89
Food Production	87
Human Nutrition	86
Youth Development	84
Natural Resources and Environment	84
Food Marketing	70
Health Care	68
Consumer Affairs	61
Economic Development	61
Family Life and Personal Development	57
Community Services and Facilities	55
Home Gardening and Lawn Care	55
Housing	46

Although staff time on this topic has been reported under such components as Ecology, Natural Resources, and Environment, clientele are suggesting that this topic is of sufficient importance to warrant future consideration. In light of the fact that special funding that has been available for energy conservation programs is declining, Extension administrators will face the decision whether to reallocate existing resources to such programs.

The decision-making process that administrators are facing in this example of energy conservation is typical of how Extension functions. Extension redirects its programs and resources in response to changes in such factors as the state of the economy, the outbreak of a new plant disease, or changes in funding patterns. But, over the long term, these efforts to address special needs have to be balanced with other program thrusts.

Often, like a sponge, Extension takes on new responsibilities as needs arise. But at the same time, it is unclear whether other thrusts ever get dropped. If they do, they seem to fade out of existence over an extended period of time so that no one notices. In order to avoid alienating supporters of a specific program, administrators appear to be reluctant to discontinue a program. The activity level in a program may be decreased, but a program is seldom completely abolished. In a way, maybe, Extension has been too successful. Because the organization has performed well, Extension staff have been called upon to carry out more and more tasks. This expression of confidence is desirable; but the organization must identify priority topics and programs ought to reflect those thrusts.

From the Present to the Future

Findings such as those summarized in this chapter begin to describe "what is." They tell us what the purposes of the organization are, how the agency is organized, how resources are utilized, what programs or activities are performed, who is reached with the programs, what is the client's reaction to programs, and anticipated support for the organization. All of these different pieces of information help to provide an understanding of how the organization is performing at the present time. While addressing what is, they do not suggest what ought to be. Questions such as "Should Extension redefine its mission statement?" "Should it direct its program at a different target audience?" or "Should mass media methods be used much more extensively?" indicate how evaluation information can be used to assist decision makers in formulating policy issues facing the organization. Extension needs to understand "what is" before it can project where the agency "ought to be" in the future. An analysis of the present situation takes on meaning when it stimulates thought and discussion concerning future actions. Documenting what exists for purposes of accountability and compliance is only part of the evaluation task. The next step is to extrapolate trends, predict changing needs, identify policy issues, and formulate Extension's response. Such is the purpose of the final chapter.

9
Policy Issues Facing Extension

Can an organization conceived in 1914 as a way to get farmers to adopt improved agricultural practices continue to be relevant when it celebrates its 100th birthday? The widely accepted singular purpose of its early history is a thing of the past. Today, it is possible to walk into a county Extension office and obtain information on such varied topics as the latest recommended varieties of wheat, precautions against the excessive use of sodium in the human diet, assistance in the development of a community water system, the treatment of scale insects on apple trees, and materials on occupational alternatives for youth. Such substantial changes in program thrusts have brought adjustments in the structure and function of the organization. It doesn't do the same things it used to do, and they aren't done in the same way.

In the preceding chapter, we have focused primarily on "what is"—what resources are used, who is aware of the programs, who uses them, and who supports them. Now, we want to shift our attention to the policy issues of the future—to "what should be." Issues that, because of their importance, just cannot be ignored.

The Mission: Should It Be Broadly Based or Narrowly Defined?

The "Extension in the '80s" report argues for a broad flexible statement of purpose so that the organization can remain relevant and respond to the dynamics of change (USDA/NASULGC, 1983). In contrast, the National Agricultural Research and Extension Users Advisory Board calls for Extension to redirect or eliminate programs and to shift personnel so they directly serve the needs of producers of U.S. food and fiber (1982). Which shall it be? Will the organization's mission be stated in terms of general education with specific programs and clientele left undefined, or will it be very specifically defined with the target audience and appropriate subject areas clearly spelled out?

The Cooperative Extension Service is a unique agency in both purpose and structure. It is in the business of education. Extension does not distribute a tangible product or service such as loans or grants, nor does it perform a regulatory function as do many government agencies. Extension also differs from most government agencies in that its mission is not defined in terms of a specific problem or issue such as housing, unemployment, or health care. Rather, it provides educational assistance on a wide variety of subjects to voluntary participants.

However, there are critics of the appropriateness of this broad-based mission. They reflect on Extension's early history and conclude that the organization should return to its original purpose of serving farmers. Others point out that it is not realistic to try to return to the days of the past. Society, including agriculture, has changed, and one cannot merely "turn back the clock" to the agency's early days. Furthermore, it could be argued that Extension's early history was not at all as it is now being portrayed. Extension played a key role in improving agricultural production, but it also stressed improved utilization of resources within the family, personal development, improved quality of life, and the improvement of the total community through the establishment of electric cooperatives, telephone systems, and rural roads. Though the specific problems being addressed differ today, the breadth of issues has changed little over the years.

Extension has not been bounded by traditional programs. Those who see the mission of Extension as defined too broadly argue that the organization should not try to be "all things to all people," but should focus its efforts. Attempts at defining a more limited mission statement tend to stress appropriate and inappropriate activities based on subject matter and clientele. The problem with using these two criteria for limiting Extension's scope is that they decrease the organization's flexibility. The genius of Extension is its responsiveness to changing needs.

The responsiveness of the agency to locally defined needs is seen as a unique characteristic of Extension that is missing from most other governmental organizations. For example, when energy prices soared, Extension mobilized energy conservation programs for homeowners, agricultural producers, and local government officials. Without this flexibility, such needs would likely go unserved. Sixty-two percent of the 1983 Extension budget came from nonfederal sources; thus state and county representatives ought to have an important voice in determining the nature of Extension's mission. A sense of program ownership and control is closely related to funding. And, since counties and states are contributing a substantial portion of Extension's budget, they expect the agency to be responsive to their needs. Like the federal government, they too designate the use for some appropriated funds. And, inasmuch as they are generally closer to state and county staff, local and state officials have subtle ways of influencing program directions. Increasing an agency's capacity to respond to problems as defined by local citizens, implies less control emanating from state and federal sources.

Agencies tend to define their roles very specifically and respond only to those issues that fit within those limits. But local citizens don't organize their needs into nice neat groupings according to the appropriate agency to respond. Their expressions are as varied as the people. As a result, the demands placed on Extension by local residents are generally more diverse than what are formulated at the federal level.

Although all levels of government assert that their top priority is to serve the needs of the people, there is still the question of who will define the needs. A recent experience of a state administrator of the Expanded Food and Nutrition Education Program (EFNEP) is illustrative. In an appearance before a federal panel, the administrator was questioned as to how it is possible to conclude that the program addresses the needs of low-income families when program administrators had not met with representatives at the national level who purport to speak for the needs of the low income. It was then necessary for the Extension administrator to inform the reviewers of the extent of daily involvement of EFNEP staff with low-income persons and Extension's system of local advisory committees. Federal representatives were identifying the needs of low-income audiences in terms of a national advocacy group, while state and county representatives defined responsiveness in terms of the expressed needs of local residents and program participants.

Extension has not limited its programs to specific audiences but makes services available without regard to the characteristics of the individual. Some observers would like to make Extension's mission clientele based. Extension would then direct its programs specifically at farmers, rural residents, minorities, or other groups of people as specified in the mission statement. Though it is likely that certain types of people find the programs of Extension more useful than others, to define Extension's target audience on the basis of such factors as place of residence, occupation, or race is contrary to the very premise of voluntary participation. People choose to use the services of Extension because they feel that the educational information they receive is useful and important to them, not because they meet specified eligibility criteria.

Extension finds itself in a dilemma, whether to formulate a very specific mission statement that clearly defines appropriate subject matter and target audience or to maintain a more broadly worded expression of purpose. Organizations have a tendency to start out general in scope and move over time toward more specificity. It is easier to justify a very specific role to a limited audience than a diffuse role to a varied clientele. A more precise statement also allows for more straightforward accountability. Most other governmental agencies are identified with the needs of a certain clientele group or societal problem, and there is pressure for Extension to do likewise. However, Naisbitt (1982) warns that organizations that identify with specific problems run the risk of becoming obsolete in a relatively short period of time.

A good example of how rigidity led to obsolescence is what happened to the railroad industry (Naisbitt, 1982). Had industry leaders conceptualized their mission in terms of comprehensive transportation systems instead of limiting it to only railroading, it is likely that they would have fared much better. They could have created a modern transportation system by combining truck, rail, and air transport. Instead, they tied their future to a single method of transportation, and, as the importance of rail transportation slackened, the organization declined along with it.

Another example of an organization that originally had a narrowly defined purpose was the Tuberculosis Society. With a decline in the incidence of the disease tuberculosis, the survival of the charitable organization was threatened. As a result, the organization broadened its statement of purpose to include all diseases and conditions of the lungs and changed its name to the American Lung Association.

Extension could face similar problems. For example, if over the years Extension had defined its clientele as *only* farmers, then the organization might have declined as have farm numbers. Obviously, Extension has not interpreted its mission so narrowly. It has been concerned with all phases of agriculture, the family, and the community. And, as a result, it has continued to prosper.

Should Extension have a broadly defined mission statement that stresses general education with specific programs and clientele left undefined, or should Extension's mission be defined more narrowly by concentrating on the needs of agricultural producers?

We would argue that Extension needs to continue to define its purpose as education, not in specific terms such as corn production, family resource management, insect control, or youth development. With education as the central focus, one can expect that specific subjects and audiences will change over time. The organization must be allowed to adjust to changing needs or find itself slipping into obsolescence. Its responsiveness must be preserved.

Structure: Should Staff Be Specialists or Generalists?

Structurally, the Extension organization is one of a kind. It involves the cooperative agreement of three levels of government and yet is not a line agency of any of the three. It also has a network of professional staff in nearly every county in the United States, a feature that is the envy of many other agencies. States have tried many variations of staffing arrangements over the years, but the importance of the county-based program has been preserved. Although technical assistance has been added at the state and area levels, the county still remains the focus of Extension programming.

The predominant Extension staffing pattern across the U.S. is a county office with three persons who have primary responsibilities for programs in agriculture, home economics, and 4-H youth. County-level staff, in turn, depend upon state and sometimes area specialist backstopping. Although variations in the pattern exist, this staffing arrangement is surprisingly uniform throughout the nation. In light of a shrinking resource base and changing program needs, this pattern of staffing is being examined. There is more emphasis on responding to high priority program needs and decreasing work on topics seen as less important. To maintain county programs, some administrators have chosen to staff county positions at the expense of state and area specialists. Other states have assigned former county personnel multicounty responsibilities in order to provide services to the same geographical area with fewer persons. Given present budgetary constraints, it is unlikely that there will be a substantial increase in the number of Extension staff. In fact, we are likely to see declines in personnel numbers. Therefore, it will be incumbent upon administrators to allocate resources and staff time in the most effective and efficient way possible. One thing is clear, decisions on staffing arrangements are complex. At a minimum, they require a consideration of budgetary impacts, programming priorities, and the roles of county and specialist staff.

All levels of government are scrutinizing agency programs from their own vantage points, and these efforts sometime work at cross purposes. For instance, changing the assignment of staff from county to multicounty areas in the interest of improved efficiency may seem logical from a state or national perspective but may lead to decreased client involvement and support at the county level. Such efforts, in the interest in overall efficiency, may fail to appreciate the importance of a presence and visibility within the local county and the role of county staff as links into broader information networks, or that local support may be jeopardized when staff are moved from the county to the area level.

During the 40s and 50s, Extension, following the trend in agriculture, emphasized increased specialization among its staff, such that by 1983 there averaged one specialist for every three county staff. Not only has the number of state and area specialists increased, but county staff have also been encouraged to obtain advanced degrees in technical areas of agriculture and home economics. However, the trend toward increased specialization may be reversing. Although competency in subject-matter areas is still seen as important, increasingly there is the realization that county staff must have a broad background and an ability to seek out solutions to a wide array of technical problems. It is impossible for county staff to be knowledgeable in all of the technical areas of agriculture and home economics, but they can know where to obtain assistance. As Naisbitt concludes, "We are moving from the specialist who is soon obsolete to the generalist who can adapt" (1982:37).

Should Extension's staffing arrangements stress greater specialization or should more emphasis be placed on the generalist role?

Perhaps, the specialist versus generalist debate is no longer the relevant issue. In an information-based society, staff will need new skills that don't necessarily fit the stereotypic roles associated with generalists and specialists. And the availability of new technologies and information systems may lead to entirely new schemes of staffing arrangements and communication methods. Such changes could mean that staff will be expected to have a working knowledge of computers and at least minimum typing skills for operating a terminal, that expertise of researchers and specialists will not stop at state boundaries but can be readily shared throughout the nation, and that state and national data bases will replace some printed publications. We would argue that Extension's most constructive response would be to bury the generalist-specialist debate and get on with identifying the needs of these new job roles, training staff in appropriate skills, and reorganizing its delivery system to fit a new age.

Resources: What Is the Best Method of Federal Funding?

The resources for Extension originate from federal, state, and local levels of government though not necessarily in equal proportions. Extension has a total budget of about $800 million, of which 38 percent comes from the federal level, 44 percent from states, and 18 percent from local government. The overall balance among these three levels of government has remained relatively unchanged for the past 25 years, but since 1980 the federal portion has started to decline and the states' proportion has increased. In addition, the earmarked part of the federal contribution has risen from 5 percent in 1970 to one-third at the present time.

The federal partner is influencing the nature and size of the organization through the budgetary process. Not only has the federal portion of the total Extension budget begun to decline, but there is also an increased tendency to earmark a greater amount. By earmarking monies, the federal government can determine the use of the funds. Such programs as Farm Safety, Expanded Food and Nutrition, Integrated Pest Management, and Rural Development came about in this manner. These designations can strain the three-way partnership.

State directors have traditionally argued for removing the earmarking of federal dollars in favor of formula funds that allow them more flexibility in use. However, a study of funding trends indicates that the funds that have been budgeted for designated purposes have been appropriated without jeopardizing the growth in formula funds. They have been added on top of normal increases in the formula category. As a result, states have received their formula fund allocation plus additional money for designated purposes. Attempts to have earmarked monies such as rural development added to the formula category have merely resulted in a replacement of existing funds. The bottom line is that, with earmarked money, states have received larger allocations than they would likely have received solely through formula funds, and when the earmarking of funds has been removed, formula funds have not increased accordingly. Therefore, in past years, it has been in the states' best interest to allow the earmarking to remain.

Seventy percent of the federal funds allocated for Extension are distributed to states on a formula basis ($205 million in 1981). The formula used for determining a state's share has undergone two major changes since its origin in 1914. Key components of the current formula are equity among states, the state's portion of the rural population, and its portion of the farm population.[1] Each time the formula has been changed, the new formula has applied only to new monies appropriated above the base as of that time. Therefore, shifts in the distribution among states have tended to be gradual.

The assumption is that the elements of the funding formula should directly reflect the purposes of the organization. The concept of distributing some resources equally among states is rationalized on the basis that it takes at least a minimum base of support to maintain a state Extension program, no matter how many people are served. The original 1914 formula also contained the criterion of rural population, and in 1953 farm population was added. The inclusion of these elements in the formula reflects the importance attributed to rural and farm residents as intended target audiences for Extension programs.

Continually, there are questions as to the adequacy of the existing formula. Concerns about the appropriateness of the present funding formula have been expressed by the USDA Users Advisory Board and in the House of Representatives Extension Oversight Hearings. The Extension in the '80s report (USDA/NASULGC, 1983:18) reflected the same thinking when it recommended, "We believe it is time for the partners to reexamine the formula for allocating federal funds to the respective state/territory partner." Since that time, an ECOP Task Force has been charged with reviewing the existing formula and has recommended changes (Task Force, ECOP, 1983). The finding from our study that a majority of Extension clientele now live in urban areas suggests that we need to consider this urban audience as well as rural residents as a component of the funding formula. Baker (1982) has suggested that the formula ought to be changed from rural population to total population so as to provide equity of service for *all* citizens. Although there are fewer rural users than urban, one should also remember that Extension still serves a greater proportion of rural residents. And we sense that, even though congressional representatives from urban districts are interested in programs that serve their constituents, they still think it is appropriate for Extension's primary focus to be rural in nature. Other than in the case of specialized programs such as urban gardening, we have not heard policymakers calling for Extension to serve the residents of large urban centers to the same degree as rural residents. If existing clientele are not to be abandoned, substantial additional resources would be necessary to support expansion of the program in order to serve all persons equitably. If the formula is broadened to include total population, we need to remember that service to rural people continues to be a high priority of the organization.

Some observers also contend that farm population is no longer an adequate indicator of Extension's agricultural focus. Cash farm receipts is suggested as an alternative. With increased specialization and concentration in commercial agriculture, a state's farm population is not necessarily reflective of the importance of agriculture in economic terms. The rationale for including cash farm receipts in the formula is that it is seen as more accurately reflecting the state's agricultural contribution to the national economy. This indicator tends to benefit states with a greater number of large commercial farms that produce high-value products.

In contrast, farm population recognizes people as the focus of the Extension effort and tends to favor states with a larger number of farmers. From the beginning, Congress chose components that recognized

Extension's purpose as "people-oriented," not ones based on economic resources. In fact, the very establishment of the Extension Service was in direct response to the misfortunes of ill-fated farm families. There is the ongoing dilemma of whether to emphasize the people or the profit in farming.

Should Extension's federal funding formula remain unchanged, <u>or</u> should it be altered to reflect changes in the organization's environment and mission?

We have pointed out some of the problems with existing and proposed elements of the formula by which the federal government allocates resources to state Extension services. It is clear that there is a need to examine periodically the funding mechanism to insure that it accurately reflects the mission of the organization. However, any changes in the funding mechanism for Extension should be closely scrutinized to insure that they truly reflect the organizational mission. Too often such decisions are made on the basis of which states stand to gain or lose money in the process.

Environmental Changes: What Is Extension's Response?

The environment in which the Extension organization carries out its programs has changed substantially over the years. The trends of yesterday are not the trends of today and will be even different tomorrow. A condensed history of the U.S. since the turn of the century might be described in the following stages: an emphasis on food production during World War I, the adjustment to a peacetime economy in the 20s, the great depression of the 30s, the all-out war effort of the 40s, the post-war technological development of the 50s, and the period of social unrest in the 60s and 70s. The commitment to social action programs of the 70s has given way to an emphasis on employable skills, a reduced role of government, and economic recovery in the 80s. From all indications, the 90s will bring even greater emphasis on information processing and utilization.

During each of these periods of American history, Extension adjusted its programs to the needs of the nation. The Extension Service played an important role in the implementation of the New Deal programs of the Roosevelt era, helped to bring agriculture into the technological age during the 50s, and emphasized disadvantaged and minority groups during the "great society" programs of the Johnson and Kennedy administrations. A new year will bring new and different demands, and it will be incumbent upon Extension to adjust accordingly.

The U.S. has moved from an industrially based society to what has been described as an information society. Rather than capital, information now is seen as the key to productivity (Naisbitt, 1982). Knowledge is viewed as the central resource of production (Drucker, 1980). However, the problem is we are about to be buried in our own information. At the present rate of growth, the quantity of scientific and technical information is doubling every five years (Naisbitt, 1982). As an information link to research findings, national computer data banks, and other telecommunication systems, Extension staff can play an important role in helping users sort through these mounds of information in order to select what fits their needs. There is no lack of information, but there is a shortage of assistance on where and how to get the right information.

The United States has also experienced a number of important population changes over Extension's 70-year history that have substantially altered its clientele base. Regional shifts have distributed more people in the West and in the sunbelt of the South and Southwest. The process of urbanization resulted in large concentrations of people in metropolitan centers and the depopulation of rural areas. Faced with the problems of population concentration and congestion, people then moved to the suburbs. And, more recently, there has begun a trend to locate in rural areas. Even though seven to ten persons still live in cities, the population is growing faster in the country.

During Extension's lifetime, the farm population has declined from a third of the total population to less than three percent. Today, "Much of the agricultural production is concentrated in the hands of relatively few producers, and most of the farm population is characterized by relatively low levels of agricultural product sales" (Stockdale, 1982:322). Less than one-fifth of the farms produce nearly four-fifths of the output. These farms are becoming larger, more specialized, more integrated, more capital intensive, and more reliant upon technological innovations. County Extension staff have a difficult time staying ahead of these specialized farmers, because such farmers often have more training and information on their particular commodity than do county staff. However, Extension staff provide a critical information linkage with university specialists and researchers.

About 80 percent of all farms could be generally described as relatively small, subsistence, part-time, retirement, and hobby farms. For many individuals, farming is as much a way of life as it is an economic enterprise. Farms and rural areas are seen as desirable places to live, own a parcel of land, and raise a family; and, when combined with off-farm employment, they provide an adequate family income. Part-time farmers, thus, are becoming the norm rather than the exception.

Many of these individuals are pursuing goals other than that of obtaining maximum productive capacity from their farms.

Should Extension strive to maintain maximum flexibility to respond to emerging needs, or should it concentrate its resources in a few areas that remain stable over time?

If Extension is going to attempt to serve the most pressing needs of a changing society, then change within the organization is inevitable. However, if Extension clings to traditional roles and refuses to change, it will be outdated in a relatively short period of time. There is a tendency within any organization to resist change and to continue to operate as in the past. Whenever an organization focuses its attention on certain issues or clientele, it sacrifices some of its ability to respond quickly to others. Adaptability is sacrificed for good performance (Weick, 1977). On the other hand, Extension could remain in a constant state of readiness and be unprepared to handle daily needs. In other words, there is a tradeoff between an organization's ability to stay flexible and its performance on specific programs. Our conclusion is that Extension has tended to stress performance over flexibility and may be performing well in some areas that are outdated. The challenge is to determine the proper balance between staying flexible and thus, responsive to changing needs, while at the same time, carrying out a quality program.

The Images of Extension: Can Extension Afford Multiple Identities?

Extension should be concerned with its image. The organization is like the proverbial elephant in the story of the blindmen's exposure to the different parts of the animal. Because of the diversity of the programs of the organization, Extension represents different things to different people. It is known as a youth group, an agency that assists farmers, a homemakers' group, a representative of the state university, and the office where you get soil tested.

These multiple identities are assets when viewed from the perspective of a single program, because people tend to make greater use of programs when they better understand the specific nature of the services offered. However, problems arise when one attempts to consolidate such diverse programs and audiences in order to develop support for the overall agency. Private sector corporations such as EXXON, IBM, and Kodak are very sensitive to the need for a single public image, so much so that they spend millions of dollars shaping it. The Extension Service, in contrast, has essentially ignored its overall organizational image and has instead chosen to promote single program identities. What one finds, as a result, is a conglomerate of divergent interest groups with little in common. In fact, there has developed an atmosphere of competition among them as they vie for the same organizational resources.

The predominant message that has been formally communicated to congressional representatives, governmental officials, and other policy makers is that Extension is an agency that serves agricultural producers. In other words, Extension has been represented almost exclusively as an agricultural agency. And, yet, that image is not reflective of the distribution of resources of current Extension programs nor the public's perception of the agency. The majority of Extension resources are devoted to programs in home economics, 4-H, and community development, not agriculture as has often been suggested.

There is no question that Extension suffers from multiple identities. 4-H has a widespread reputation as a program that teaches young people practical skills and contributes to their personal development. And, as a program title, 4-H is the most widely recognized Extension program. However, the name recognition of the 4-H program has not been transferred to the Extension organization. Organizational staff could be faulted for not stressing that linkage. Like 4-H, the home economics, community development, and agriculture and natural resources programs deal with distinct subject matters and have loyal followers. Unfortunately, these supporters are individuals with very different interests, and bringing them together on behalf of the organization is not an easy task. Nevertheless, it is an important one.

Should Extension continue to maintain multiple images that center around its distinct programs, or should it strive to mold a single unified identity?

Extension is a single organization performing multiple functions. But that situation is not unique. Private sector organizations project a single organizational image while developing loyalty to a large number of products and services. Extension needs to develop legitimacy as a multiple-function educational organization. And with institutional legitimacy for the umbrella organization, Extension can continue to carry out a variety of projects and programs. We are not suggesting that in order to develop a unified image, Extension has to limit itself to a single function. On the contrary, the organization can continue to carry out multiple functions with a variety of audiences. However, what it does suggest is that program efforts and accomplishments be clearly identified with the Extension organization. The future of Extension may depend upon the successful molding of this single image.

Urban and Rural Clientele: Who Should They Be?

Over eleven million households report using the services of Extension in a year (1981) and twenty-two million have used them sometime during their lifetime. In orther words, over a quarter of the U.S. population have at one time or another made use of the services Extension offers. Without getting into a discussion of whether that estimate is high or low, when it is compared with the number of clientele of other agencies, it must be concluded that this number represents a fairly high level of program penetration. But should it be reaching more people? There is an implicit assumption that the agency should continue to increase the number of its clientele, though such an inference is not necessarily based on any analysis of fact. Rather, the assumption is the more people reached, the better. But when is enough enough? There have been few, if any, efforts to balance the quality of the experience with quantitative measures of participation or to identify limits on the appropriate number of clientele.

To a great extent, the households that report being users of Extension services are typical of the total U.S. population. Most live in urban areas and have about an average family income and educational level. One can accurately conclude that Extension predominately serves middle Americans. However, that conclusion is not necessarily meant as a criticism. It merely recognizes the fact that, as a voluntary educational program, Extension is attracting a cross section of the public. This finding should be seen as negative only if one concludes that Extension ought to be serving a specific audience other than the general population. Because historically the organization focused on the provision of educational information on agriculture and rural living and Extension was organizationally established in the U.S. Department of Agriculture and colleges of agriculture, it is not surprising that some observers have concluded that the organization ought to serve primarily farm and rural people. While serving more urban dwellers, Extension still reaches a greater proportion of rural and farm residents. Extension does not exclusively serve rural people; it never did, but people are more likely to avail themselves of the agency's programs if they live in rural areas or on a farm. Thus, service to rural and farm residents is still fundamental to the organization.

Should Extension define its target audience as rural and farm residents, or should its intended clientele be left undefined?

The issue here is whether it is appropriate to define Extension's purpose in terms of the rural and urban residency of the target audience. Or, more accurately, will Extension develop programs to speak to the needs of only rural residents and not those of urban people? In many respects, it would be easier if the audience was clearly identified, then accountability would be much simpler. However, the basic principle of offering equal access to the educational programs of the agency would be violated. Such a move would also ignore the reality that Extension is presently serving more urban than rural people. The organization cannot ignore that fact and decide to turn its back on the majority of its clientele. We submit that short of abandoning existing clientele, there really is no viable alternative of Extension limiting its services only to rural and farm residents.

Equal Opportunity:
How Do the Disadvantaged and Minorities Fare?

People do not feel discriminated against by Extension. Only five persons expressed any type of overt discrimination, and part of these complaints were unfounded. However, even though there is a feeling of equal opportunity among existing users, Extension reaches a lower proportion of blacks and Hispanics than is present in the total population. A notable exception is the community development program, in which a much greater proportion of racial minorities participate.

Some observers contend that Extension should serve a greater number of the disadvantaged and minorities and fewer high income persons. There is no question that the nature of a program influences, at least to some extent, the makeup of its clientele. If efforts such as the Expanded Food and Nutrition Education Program and urban gardening are major components of Extension programs, then low-income persons are more likely to comprise a larger portion of its clientele.

Disadvantaged persons often need more personal attention and counseling than do persons of higher income and educational levels. As a result, such programs require more staff time and resources per person assisted. Yet, in the face of tight budgets, there is pressure to increase the use of mass media methods. In the interest of reaching larger numbers of clientele more efficiently, programs for audiences with special needs could suffer.

Should Extension give increased attention to the services provided the disadvantaged and minorities, or should the composition of Extension clientele be left to that resulting from voluntary participation?

In addition to equal opportunities according to the criteria of race, sex, and age, there are other subtle ways in which Extension may be discriminating. Agricultural advisories have tended to have a pro-commerical farmer bias, home economics programs have emphasized the needs of families, 4-H activities and projects generally require at least a minimum level of monetary resources, and community development stresses the community self-help approach. All of these assumptions cause Extension's programs to be less useful and appropriate to some people than others. To the small subsistence farmer, the single parent, the low-income youth, and the public housing resident, the programs of Extension may seem discriminatory. Extension needs to be particularly mindful of the biases and assumptions built into programming decisions that lead to covert discrimination.

However, even with programs directed toward specific clientele groups, it must be remembered that Extension has no captive audience. People are free to use or ignore the services offered. In the end, it will be the recipients themselves who will determine the composition of agency clientele.

Use: Try It, You'll Like It

Like the line in the popular television commercial, when they try it, clients really do like the service they receive from Extension. Whether they are rich or poor, young or old, black or white, clientele are satisfied with Extension. It has often been said that Extension operates a quality program that is developed in response to locally expressed needs, such that people feel it is their program. Evidently, there is truth in that statement because the users of the service express a very high level of satisfaction with Extension, greater than that found in the assessment of other public agencies.

The nature of the Extension program is at least one factor that contributes to its good rating. The organization offers voluntary educational programs that are not threatening to the participants. People tend to seek out Extension for assistance, and, in doing so, they make a commitment to the solution of a problem. This causes people to be much more receptive to the information provided. And, because Extension's product is education there is no clear differentiation of clients on the basis of eligibility requirements or services rendered such as there is with public assistance, loan programs, and the like.

Compared with most other governmental agencies, Extension offers a greater opportunity for local residents to influence program directions. The organization engages in an involved process of program planning that starts with expressions of need from within the community and culminates in programs in response to the problems identified. This program planning process is based on the premise that ideas for program directions ought to originate from a set of circumstances at the local level rather than as strategies imposed from above. Accurate expressions of local needs are dependent, however, upon involvement that is representative of all segments of the community. There is always the danger that, if county advisory committees are composed of traditional Extension supporters, then new program emphases will be overlooked. In which case, Extension will find itself in a position of supporting the status quo and attracting few new clientele.

Should Extension develop programs based on the needs as expressed by the present users of Extension services, or should it strive to program in response to an assessment of needs of all residents?

As has been reported, people who use Extension are overwhelmingly satisfied with the service they receive. But what kinds of impressions do the nonusers hold? What is the perception of the other three-fourths of the population? Most of these people have either heard of the agency name or one of its programs but have never had any personal experience with it. As a result, about half of these persons have not formed an opinion of Extension. They lack sufficient information from which to judge and, thus, are undecided. Inasmuch as use is so closely related to satisfaction, if these individuals become Extension users, they would also be expected to give the agency a positive rating. So, if use is increased, satisfaction could be expected to do likewise.

In addition, Extension ought to focus on the expression of the slightly more than half who are nonusers, but *have* formed an opinion about Extension. This group can provide important information as to why they have chosen to be nonusers. They can also furnish insight into what has been referred to as the "common stock of knowledge" that comprises Extension's organizational image (Berger and Luckman, 1967). This group has generally been ignored in past evaluation studies, but because they also contribute tax dollars in Extension's support, they should be included in future evaluation efforts. We would argue that Extension must recognize that its support comes from all taxpayers, not just those who use the service; thus, the future of the organization may rest as much with nonusers as users.

Extension Methods: Personal or Impersonal?

County Extension staff communicate with clientele through three main methods: individual contact, group meetings, and mass media. Throughout the history of the organization, there was an emphasis on the use of demonstrations and home and farm visits as appropriate Extension methods. However, in more recent years, there has been a shift away from individual methods to an increased reliance upon group and mass media methods. This trend has been encouraged by the development of more sophisticated telecommunication systems. And, yet, even with the availability of such innovations as television, teleconferencing, videotapes, videotext, and electronic messaging, slightly more than half of the time county agents in Kentucky spent in contact with clientele was through the use of individual methods. And personal visits by agents at the client's location represented 39 percent of the total.

Individual conferences with clients, especially at their location, are time consuming and expensive. Some critics would conclude that spending half of the professional staff time in personal contact with clientele is too much. They would argue that at least some of the information imparted in an individual visit can just as effectively be communicated through much cheaper methods. Others contend that it may be possible to transmit the information to a greater number of people, more cheaply, and quicker through the mass media, but clients may be unable to adapt it for use in their particular situations. A telecommunications system may be more efficient in terms of messages delivered but may be very ineffective in terms of stimulating behavioral changes. What is the proper balance among individual, group, and mass media methods? No one seems to know. Most would agree that it takes a combination of methods to communicate effectively and efficiently but are unsure what the mix ought to be.

It is generally recognized that Extension communicates different types of information to different audiences. The information may be of a simple factual nature, or it may be very complex. It may be used as input in making a major decision that requires considerable counseling, thought, and contemplation. Or it may be something as straightforward as the proper rate of application of a pesticide. In some cases the intended recipient of the information may be well educated and able to assimilate the information without assistance, and in others, the client will require considerable personal attention. This discussion suggests that Extension needs to examine the appropriateness of different communication methods, with an eye toward making them both more efficient and effective. However, it is important to remember that Extension's mission is education, not just information transmission. Also, information needs to be repackaged for use in more than one media. For example, with a minimum amount of extra effort, a workshop topic can become a newspaper or magazine article or the subject for a radio or television broadcast. Combinations of methods are likely to be more successful.

Will Extension recognize the changes occurring in an information society and make the necessary adjustments in information delivery methods, or will it choose to ignore the changes taking place and proceed with business as usual?

As an educational organization, Extension is in the business of creating, processing, and distributing information. That places the agency in a strategic position to assist society in moving into the information age. However, Extension's ability to respond will depend, to a great extent, upon the organization's own response to change. Extension must consider using more innovative approaches to information processing and distribution such as computers for data storage and manipulation, word processing, national and international data bases, selective retrieval of information, and telecommunications systems that reduce "float time" of transmissions. Extension, as an organization, needs to sort out these issues before it can provide meaningful assistance to clientele. As new telecommunication systems become available, we contend that Extension needs to assess their proper role in the organization's educational program and, when found beneficial, to be a leader in their adoption and use.

Support: Who Will Champion Extension's Cause?

There is public consensus that support for Extension jshould be at least as great as it is now. Almost no one wants it reduced. Users are more supportive than nonusers, support increases as the frequency of use increases, and the more satisfied people are with Extension the more they want it supported. Therefore, support can be improved by increasing the number of satisfied users. Even among persons who have not used Extension, the organization has a positive enough image that people are willing to see the present level of support maintained.

The expression of support reported in this study is an indicator of the level of confidence that individual citizens place in the Cooperative Extension Service. It registers a general sentiment that is a more specific statement than merely indicating a general level of satisfaction or dissatisfaction with a service. Desired funding support is a "pocketbook" issue in that it forces each person to take a stand on the allocation of public resources.

However, it must be remembered that this survey response is a compilation of individual views. And, because of the reality of the decision-making process, political mechanisms tend to be more responsive to organized expressions. Individual consumers of the service can exert pressure on political authorities, although it is likely that they will have limited influence on those who are in control of the process of resource allocation. Organized efforts tend to have more impact. But the question is, "How is public support related to organized efforts?"

Public support is seen as a necessary but not sufficient condition for organized support. The expectation that individuals will rally to the aid of an agency in times of need may be overly optimistic. Most consumers of a service have a fragmented and limited interest in a specific public agency and are seldom willing to devote time and energy to insuring its survival and growth. Though public sentiment is an important component of organizational support, the task of speaking on behalf of organizations usually falls to organized, ongoing special interest groups. Their success is at least partially governed by their relative influence among decision makers and the counterpressures being exerted by other special interest groups that are competing for the same resources.

Over the years, the Extension Service has depended primarily upon agricultural producers and farm organizations to represent the interests of Extension among policymakers at the federal level. As evidenced by the growth in the Extension budget over the years, this strategy has proved to be successful. However, Extension programs are broader than just agriculture, and, in addition, the political organization of agriculture has moved from purposes generally supported by the general public to narrowly defined speical interest legislation (Bonnen, 1965). Agricultural policy has come to mean legislation dealing with specific commodities such that it has become increasingly difficult to develop broad-based policies. The farm bloc of the 1920s has splintered into the special interests of such commodities as wheat, cotton, tobacco, and peanuts, all of which are competing with each other for improved commodity programs. At best, one could conclude that there exists a shaky policy-making coalition within agriculture.

As a minority rather than a majority, rural and farm interests must look to alliances with other groups such as labor, consumers, and urban constituencies. Passage of rural and farm legislation will depend, to a great extent, upon strategies of vote trading and coalitions with nonfarm interests. Problems of farm and rural people will increasingly need to be stated in terms of their impact on consumers and urban residents.

Should Extension continue to rely primarily on agricultural support groups to represent its interests, or should it develop a support base that more nearly reflects the breadth of its clientele?

Extension has continued to rely almost exclusively upon agricultural interests to be its voice in Washington D.C., to the extent that one could conclude that, as agriculture goes, so goes Extension. At the same time, nonagricultural interests have become more prevalent in Extension program offerings and rural and farm political influence has diminished.

The results of this study have demonstrated that support is present among nontraditional audiences, but it is up to Extension to decide whether it wants to broaden its support base, and if so, how it would be done. For example, many local officials have received the benefits of community development programs, but they seldom have been mobilized to speak on behalf of Extension at statehouses and in Washington. It is our contention that Extension needs to update its support base to reflect the true nature of its clientele and program thrusts.

Evaluation: What Is and What Ought to Be?

To date, Extension evaluation efforts have been largely piecemeal and in response to the way winds are blowing at the federal level. The organization has passed through stages that have stressed the documentation of how staff time is devoted to different program activities; "head counting" to register the number of persons contacted by program staff; administrative audits that have concentrated on compliance with equal opportunity criteria; and recently, impact studies that attempt to demonstrate changes in the recipient population that are attributable to the programs. All of these are important evaluation methods, but none is adequate by itself. This fact is evident in the "Report of the National Task Force on Extension Accountability and Evaluation System" (1981) in which the members call for a comprehensive evaluation system that includes impact studies, accomplishment information, and input information.

Hopefully, this book will stimulate people to think of evaluation from a total systems perspective that follows the elements of the programming process. There are certain *inputs* that are converted into *activities* that then have an *impact* on peoples' lives within a larger *environment*. In the Systems Effectiveness Model we have suggested what are some of the components for determining the effectiveness of Extension, as well as how they are interrelated. Still, it provides only a beginning. Extension has barely scratched the surface of the many alternative approaches and data sources available. Staff members spend a considerable amount of time accumulating mounds and mounds of information, and yet little use is made of it. And there have been almost no attempts to integrate different sources of data in order to provide a more comprehensive picture.

The point is that, before Extension becomes so concerned with new evaluation methods, it needs to do a much better job of thinking through what constitutes the inputs, activities, and impacts of a program and then using evaluation information that is already available.

Goal-oriented accountability systems like EMIS rely primarily on statements of accomplishment by agency personnel. As a result, constituents, either as direct recipients of the services or as cost-bearers, have been largely ignored. In contrast, the approach used in this study relies heavily upon the expression of constituents. As a result, the intended recipients of the programs are the persons deciding upon appropriate criteria for assessing effectiveness of performance. In this context, Extension will be judged as effective if it is perceived as meeting the needs of constituents. A critical concern for evaluation efforts is whether constituents rate the performance differently than do agency personnel.

The emphasis on client-based evaluation is not intended as a replacement for assessments by agency personnel or impact studies that document the social and economic value of certain programs. Rather, "perception data" should be used along with agency-generated indicators and value-added measures. The end product should then be a much more complete evaluation that considers that the people who are supposedly being served by an agency have experiences and reactions that are important in the assessment of performance of that agency.

Because Extension's product is educational in nature, it is often difficult to account for the resources invested in the effort. Accomplishments are not easily quantified and cause-and-effect relationships are unclear. It is also extremely difficult to separate the impacts attributable to Extension programs. As a result, Extension will continue to struggle to identify acceptable indicators of organizational effectiveness. Even with tests of achievement within controlled environments, the accomplishments of school systems are still being questioned. Therefore, even though more sophisticated evaluation techniques are needed in Extension, it cannot be expected that their results will satisfy all of the need to be accountable. Organizational evaluation poses a series of complex questions for which we have only limited answers.

Will the current emphasis on evaluation result in the utilization of findings for improved programming, or will Extension ignore the results and continue with traditional methods?

Evaluation studies will not solve all of the problems of Extension. On the one hand, policymakers complain that evaluation efforts are not providing responses to the questions they need answered; while, on the other hand, agency staff are frustrated because they feel the information they are providing is being ignored. Evaluation results provide information for decision makers, who, in turn, must integrate the evaluation results with information from other sources. To the frustration of many evaluators, final policy decisions are often contrary to the recommendations of evaluations. However, to suggest that evaluation results ought to directly determine final decisions is to ignore the political reality of public policy. Final decisions are made after weighing the evidence from a multitude of sources; only one of which is the evaluation study. From the standpoint of evaluators, traditional methods, common sense, and political influences may not be seen as rational bases for determining program directions, but they will continue to be important elements of public policy decisions. What is hoped is that over time there will develop a climate of greater receptivity to the utilization of evaluation findings by policymakers. However, a large share of the responsibility for increasing the utilization of evaluation results lies with agency staff and evaluators (Patton, 1978). By involving decision makers and information users in the design and implementation of evaluation studies, utilization is then enhanced. However, it is our opinion that the only way that meaningful evaluation will occur, is if Extension makes a serious commitment to accountability and evaluation efforts by being willing to devote adequate staff time and resources to the implementation of evaluation plans and by seeing that results are utilized in programming.

The Book's Intent

The goal of this book was to increase our understanding of the nature of organizational effectiveness. The model that is presented recognizes the major elements and processes of the organization within its environment. It is anticipated that a better understanding of these components and their interrelationships will lead to a greater likelihood of organizational success.

This effort has viewed the total Extension organization in its sociopolitical environment. We have tried to raise and explore some of the critical policy issues facing Extension today and to move the policy discussion from documenting "what is" to suggesting "what ought to be." Extension is drifting in the winds of conflicting expectations and changes in resource allocation. Policy considerations are needed to rationally chart a course that will maximize the strengths of the organization and enhance its relevance to an evolving society.

This book has been written with the deep conviction that major changes are occurring in the United states as it moves from an industrial to an information society, changes unlike any that have been experienced before. Extension is vastly different than it was seventy years ago and, yet, we can expect the organization to undergo even greater transformations in the years ahead. The question is, "What will be Extension's response to these changes?" Extension has been and continues to be an important information agency and stands at a crossroads in this evolving age. Either Extension can anticipate such changes and be an important agent of change, or it can ignore them and be dragged "kicking and screaming" into the information age. Extension can shape its own destiny, or it can allow its future to be molded by others.

Notes

1. The exact formula is as follows: 4% to Federal Extension Service and 96% to states as follows: 20% in equal portions, 40% on basis of rural population, and 40% on basis of farm population.

References

Andrews, Frank, and Stephen Withey
    1976 Social Indicators of Well-Being. New York: Plenum.
Baker, Ron
    1982 Distribution of Smith Lever Funds: Change Is Essential. Unpublished Mimeo. Berkeley: University of California.
Ballew, Ralph J., et al.
    1976 "In service to agriculture." Pp. 32–49 in C. Austin Vines and Marvin A. Anderson (eds.), Heritage Horizons: Extension's Commitment to People. Madison, Wisconsin: Journal of Extension.
Bennett, Claude F.
    1982 "Evaluating joint effects of Extension programs: toward integration in the social sciences." Rural Sociologist 2:163–172.
Berger, Peter, and Thomas Luckman
    1967 The Social Construction of Reality. Garden City, New Jersey: Doubleday.
Bonjean, Charles M., Harley L. Browning, and Lewis F. Carter
    1969 "Toward comparative community research: a factor analysis of United States counties." Sociological Quarterly 10:157–176.
Bonnen, James T.
    1965 "Present and prospective policy problems of U.S. agriculture: As viewed by an economist." Journal of Farm Economics 47:1116–1129.
Brickman, Philip, and Donald T. Campbell
    1971 "Hedonic relativism and planning the good society." In M. H. Appley (ed.), Adaptation-Level Theory: A Symposium. New York: Academic Press.
Cameron, Kim
    1981 "The enigma of organizational effectiveness." Pp. 1–13 in Dan Baugher (ed.), Measuring Effectiveness. San Francisco: Jossey-Bass Inc.
Campbell, Angus, and Phillip E. Converse
    1972 The Human Meaning of Social Change. New York: Russell Sage Foundation.
Campbell, Angus, Phillip Converse, and Willard Rogers
    1976 The Quality of American Life. New York: Russell Sage Foundation.

Campbell, Charles, Larry Hall, and B. W. Harrison
  1971 "A study of Extension program planning as perceived by off-campus faculty, lay leaders and the general public in the show-me area." Columbia: University of Missouri.
Christenson, James A.
  1976 "Quality of community services: a macro-unidimensional approach with experiential data." Rural Sociology 41:509–525.
Christenson, James A., and Gregory S. Taylor
  1982 "Determinants, expenditures, and performance of common public services." Rural Sociology 47:147–163.
Christenson, James A., and Paul D. Warner
  1982 "An assessment model for the Cooperative Extension Service." Rural Sociology 47:369–390.
Christenson, James A., Paul D. Warner, and JoAnne V. Day
  1980 "Rural-urban services in Kentucky." Community Development Issues 2. Lexington: University of Kentucky.
Cosner, Barney L., C. Wesley Holley, Thomas Randle, Eddy Finley, and James P. Key
  1980 "The awareness of the general public of Oklahoma of the instruction, extension, and research components of the division of agriculture at Oklahoma State University." Unpublished report. Stillwater: Oklahoma State University.
Coughenour, C. Milton, Ann Stockham, and James A. Christenson
  1980 "Kentucky Farm Families." Community Development Issues 2 (5). Lexington: Department of Sociology, University of Kentucky.
Cronback, Lee J., and Associates
  1980 Toward Reform of Program Evaluation. San Francisco: Jossey-Bass Publishers.
Cummings, Larry L.
  1977 "Emergence of the instrumental organization." Pp. 56–62 in Paul S. Goodman, Johannes M. Pennings, and Associates (eds.), New Perspectives on Organizational Effectiveness. San Francisco: Jossey-Bass Publishers.
Dahl, Robert A.
  1956 A Preface to Democratic Theory. Chicago: University of Chicago Press.
DeMarco, Susan
  1980 "County visits." In Evaluation of Economic and Social Consequences of Cooperative Extension Programs. Appendix I. Washington, D.C.: U.S. Department of Agriculture.
Development Planning Associates
  1977 Evaluation of Nutrition Planning Workshops. Washington, D.C.
Drucker, Peter
  1980 Managing in Turbulent Times. New York: Harper and Row.
Edelman, Murray
  1971 Politics as Symbolic Action: Mass Arousal and Quiescence. Chicago: Marham Publishing.

References

Etzioni, Amitai
  1964 Modern Organizations. Englewood Cliffs, New Jersey: Prentice-Hall.
  1968 The Active Society. New York: The Free Press.
Friedlander, Friedrich and Hal Pickle
  1968 "Components of effectiveness in small organizations." Administrative Science Quarterly 13:289–304.
Fuguitt, Glen
  1965 "Career patterns of part-time farmers and their contact with the Agricultural Extension Service." Rural Sociology 30:49–62.
Gallup
  1979a "The Gallup study of adults' and childrens' participation in 4-H youth programs." Princeton, New Jersey: The Gallup Organization, GO 78186.
  1979b "The Gallup study of participation and awareness of the Agricultural Extension Service." Princeton, New Jersey: The Gallup Organization, GO 7925.
Georgopoulos, Basil S., and Floyd C. Mann
  1962 The Community General Hospital. New York: Macmillan.
Georgopoulos, Basil S., and Arnold S. Tannenbaum
  1957 "A study of organizational effectiveness." American Sociological Review 22:534–540.
Groves, Robert M.
  1978 "An empirical comparison of two telephone sample designs." Journal of Marketing Research 15:622–631.
Hasenfeld, Yeheskel, and R. A. English
  1974 Human Service Organization. Ann Arbor: University of Michigan Press.
Hildreth, R. J., and Walter J. Armbruster
  1981 "Extension program delivery—past, present, and future: an overview." American Journal of Agricultural Economics 63:853–858.
Holzer, Marc (ed.)
  1976 Productivity in Public Organization. New York: Kennikat Press.
Janowitz, Morris, Deil Wright, and William Delany
  1958 Public administration and the Public, Ann Arbor: University of Michigan.
Jenkins, John W.
  1980 "Historical overview of Extension." In Evaluation of Economic and Social Consequences of Cooperative Extension Programs. Appendix I. Washington, D.C.: U.S. Department of Agriculture.
Kahn, Robert L.
  1977 "Organizational effectiveness: an overview." Pp. 235–248 in Paul S. Goodman, Johannes M. Pennings, and Associates (eds.), New Perspectives on Organizational Effectiveness. San Francisco: Jossey-Bass Publishers.
Kappa Systems Inc.
  1979 Extension Program Impact Findings from Selected Studies Conducted from 1961–1978. Volume II. Arlington, Virginia.
Katz, Daniel
  1960 "The Functional Approach to the Study of Attitudes." Public Opinion Quarterly 24:163–204.

Katz, Daniel, Barbara Gutek, Robert Kahn, and Eugenie Barton
   1977 Bureaucratic Encounters. Ann Arbor: Institute for Social Research.
Katz, Daniel, and Robert L. Kahn
   1978 The Social Psychology of Organizations. Second Edition. New York: John Wiley and Sons.
Katz, Elihu, and Paul F. Lazarsfeld
   1955 Personal Influence: The Part Played by People in the Flow of Mass Communication. New York: Free Press.
Kelling, G., T. Pate, D. Dieckman, and C. E. Brown
   1974 The Kansas City Preventive Patrol Experiment: A Technical Report. Washington: Police Foundation.
Klapper, Joseph T.
   1960 The Effect of Mass Communication. New York: Free Press.
Leff, Herbert L.
   1978 Experience, Environment, and Human Potentials. New York: Oxford University Press.
Loomis, Charles P.
   1953 Rural Social Systems and Adult Education. East Lansing: Michigan State University Press.
Lyons, Larry
   1977 "Community power and policy outputs: a question of relevance?" Pp. 418–434 in Roland Warren (ed.), New Perspectives on Community in America. Chicago: Rand-McNally.
MacIver, Robert M.
   1955 Academic Freedom in Our Time. New York: Columbia University Press.
Marans, Robert W., and Willard Rodgers
   1975 "Toward an understanding of community satisfaction." Pp. 299–352 in Amos Hawley and Vincent Rock (eds.), Metropolitan America in Contemporary Perspective. New York: John Wiley and Sons.
Maslow, Abraham H.
   1971 The Farther Reaches of Human Nature. New York: Viking.
Meier, Kenneth J., and William P. Browne
   1983 "Interest groups and farm structure." Pp. 47–56 in David E. Brewster, Wayne D. Rasmussen, and Garth Youngberg (eds.), Farms in Transition. Ames: Iowa State University Press.
Mendelsohn, Harold
   1975 "Some Reasons Why Information Campaigns Can Succeed." Pp. 304–315 in Susan Welch and John Comer (eds.), Public Opinion. Palo Alto, California: Mayfield Publishing Company.
Meyer, Marshall W., and Associates
   1978 Environments and Organizations. San Francisco: Jossey-Bass Publishers.
Miller, Robert W.
   1979 Evaluative Research in Rural Development: Concepts, Methods, and Issues. Ithaca, New York: Cornell University.
Moos, Rudolf
   1974 Evaluating Treatment Environments. New York: Wiley-Interscience.

References

Mulford, Charles L., Gerald E. Klonglan, Richard D. Warren, and Ronald C. Powers
  1980 "Impact of Extension's community resource development projects: a study involving state program leaders, Extension workers, and knowledgeable citizens." Sociology Report No. 146. Ames: Iowa State University.
Naisbitt, John
  1982 Megatrends. New York: Warner Books.
National Agricultural Research and Extension Users Advisory Board
  1982 Agricultural Research, Extension, and Education Recommendations. U.S. Department of Agriculture, Washington, D.C.
National Task Force on Extension Accountability and Evaluation System
  1981 Report of the National Task Force on Extension Accountability and Evaluation System. Morgantown: West Virginia University, Cooperative Extension Service.
Nolan, Michael, and Paul Lasley
  1970 "Agricultural Extension: Who uses it?" Journal of Extension 17:21–27.
Odiorne, George S.
  1965 Management by Objectives: A System of Managerial Leadership. New York: Pitman.
Patton, Michael W.
  1978 Utilization-Focused Evaluation. Beverly Hills, California: Sage Publications.
Pennings, Johannes, and Paul S. Goodman
  1977 "Toward a workable framework." Pp. 146–184 in Paul S. Goodman and Johannes Pennings (eds.), New Perspectives on Organizational Effectiveness. San Francisco: Jossey-Bass Publishers.
Perrow, Charles
  1970 Organizational Analysis: A Sociological View. Belmont, California: Wadsworth Publishing Co.
Peters, John G.
  1978 "The 1977 farm bill: coalitions in Congress." Pp. 23–35 in Don F. Hadwiger and William P. Brown (eds.), The New Politics of Food. Lexington, Massachusetts: D. C. Heath.
Pigg, Kenneth E., and James M. Meyers
  1980 Social and Economic Consequences of the 4-H Program. Volume 1. Washington, D.C.: Extension Service, United States Department of Agriculture.
Price, James L.
  1968 Organizational Effectiveness: An Inventory of Propositions. Homewood, Illinois: Richard D. Irwin, Inc.
Price, James
  1972 "The study of organizational effectiveness." Sociological Quarterly 13:3–15.
Provus, Malcolm
  1971 Discrepancy Evaluation for Educational Program Improvement and Assessment. Berkeley, California: McCutchan.

Rossi, Peter H., Richard A. Berk, and Bettye K. Eidson
  1974 Roots of Urban Discontent. New York: John Wiley.
Rossi, Peter, Howard Freeman, and Sonia Wright.
  1979 Evaluation: A Systematic Approach. Beverly Hills, California: Sage Publications.
Scott, W. Richard
  1977 "Effectiveness of organizational effectiveness studies." Pp. 63–95 in Paul S. Goodman, Johannes M. Pennings, and Associates (eds.), New Perspectives on Organizational Effectiveness. San Francisco: Jossey-Bass Publishers.
Scott, W. Richard, et al.
  1978 "Organizational effectiveness and the quality of surgical care in hospitals." Pp. 290–305 in Marshall W. Meyer and Associates, Environments and Organizations. San Francisco: Jossey-Bass Publishers.
Shin, Don C.
  1977 "The quality of municipal service: concept, measurement and results." Social Indicators Research 4:207–229.
Smith, Clarence B., and Meredith C. Wilson
  1930 The Agricultural Extension System of the United States. New York: John Wiley and Sons.
Stagner, Ross
  1970 "Perceptions, aspirations, frustrations and satisfactions: an approach to urban indicators." The Annals of the American Academy 388:59–68.
Steers, Richard
  1977 Organizational Effectiveness: A Behavioral View. Santa Monica, California: Goodyear Publishing Co.
Stockdale, Jerry D.
  1982 "Who will speak for agriculture." Pp. 317–327 in Don Dillman and Daryl Hobbs (eds.), Rural Society in the U.S. Boulder, Colorado: Westview Press.
Suchman, Edward A.
  1967 Evaluative Research. New York: Russell Sage Foundation.
Summers, James C., Robert W. Miller, Cecil E. Carter, Richard E. Young, Carolyn Demsey-Foss, and Lee Beaumont
  1981 Program Evaluation in Extension: A Comprehensive Study of Methods, Practices, and Procedures. Morgantown: Center for Extension and Continuing Education, West Virginia University.
Tanner, Bonnie, June Bryan, and Kathy Rygasewicz
  1982 State Survey: Redirection of Program Efforts. Working Paper. Washington, D.C.: Extension Service, U.S. Department of Agriculture.
Task Force, Extension Committee on Organization and Policy
  1983 Formula for Distribution of Smith-Lever Funds to States. Pullman: Washington Cooperative Extension Service.
Thompson, James D.
  1967 Organizations in Action. New York: McGraw-Hill.

Tripodi, Tony, Phillip Fellin, and Irwin Epstein
 1971 Social Program Evaluation: Guidelines for Health, Education, and Welfare Administration. Itasca, Illinois: F. E. Peacock.
True, Alfred C.
 1928 A History of Agricultural Extension Work in the United States: 1785–1923. Washington, D.C.: U.S. Government Printing Office.
United States Department of Agriculture
 1980a Evaluation of Economic and Social Consequences of Cooperative Extension Programs. Washington, D.C.
 1980b Evaluation of Economic and Social Consequences of Cooperative Extension Programs. Appendix I. Washington, D.C.
United States General Accounting Office
 1981 Cooperative Extension Service's Mission and Federal Role Need Congressional Clarification. CED-81-119. Washington, D.C.
U.S. Congress, House Committee on Agriculture
 1913 Cooperative Agricultural Extension Work. Report No. 110. Washington, D.C.
USDA/NASULGC Joint Committee
 1983 Extension in the '80s: A Perspective for the Future of the Cooperative Extension Service. Madison: University of Wisconsin, Cooperative Extension Service.
 1968 A People and a Spirit. Fort Collins: Colorado State University.
U.S. House of Representatives
 1982 Extension Service Oversight, Serial No. 97-EEE. Washington, D.C.
 1981 "Issue Areas for Cooperative Extension System Oversight." Subcommittee on Department Operations, Research and Foreign Agriculture. Washington, D.C.
Vanfossen, Beth Ensminger
 1979 The Structure of Social Inequality. Boston: Little, Brown and Company.
Warner, Paul D., Rabel Burdge, Susan Hoffman and Gary Hammonds
 1975 "Issues Facing Kentucky." Lexington: Department of Sociology, University of Kentucky.
Warner, Paul D., and James A. Christenson
 1981 "Who is Extension serving?" Journal of Extension 19:22–28.
Warner, Paul D., and James A. Christenson
 1983 "Looking Beyond Extension Stereotypes," Journal of Extension 21:27–33.
Weick, Karl E.
 1977 "Re-punctuating the problem." Pp. 193–234 in Paul S. Goodman, Johannes M. Pennings, and Associates (eds.), New Perspectives on Organizational Effectiveness. San Francisco: Jossey-Bass Publishers.
Weiss, Carol H.
 1972 Evaluation Research: Methods of Assessing Program Effectiveness. Englewood Cliffs, New Jersey: Prentice-Hall.
Wholey, Joseph S., John W. Scranlon, Hugh G. Duffy, James S. Fukumoto, and

Leona M. Vogt
　1976 Federal Evaluation Policy: Analyzing the Effects of Public Programs. Washington, D.C.: The Urban Institute.
Wisconsin Extension Staff
　1979 "How many Wisconsin adults use Extension?" Unpublished report. Madison: University of Wisconsin-Extension.
Yuchtman, Ephraim, and Stanley Seashore
　1967 "A system resource approach to organizational effectiveness." American Sociological Review 32:891–903.

Appendix A: The National Survey

The findings of the national survey are based on 1,048 completed telephone interviews with a nationwide probability sample of the public 18 years of age and over conducted by The University of Kentucky Survey Research Center in 1982. Overall, 70 percent of potential valid respondents were interviewed.

It is difficult to calculate the response rate for a telephone survey because of the many possible outcomes of random digit dialing selection and of phone calls across the U.S. The response rate used is calculated as follows:

$$\frac{\text{Persons known eligible and interviewed}}{\text{Eligible and interviewed + refused + noncontact}}$$

The "noncontact" category includes respondents screened to be eligible but who could not be contacted after ten attempts at five different time periods. Refusals included persons identified as eligible by a screening procedure who refused to participate, plus estimates of the proportion of eligible cases where the person who answered the phone refused to participate in the screening procedure.

Sample Description

A sample of telephone numbers was purchased from the University of Michigan Survey Research Center. The following summarizes the structure of the sample. For a complete description of the sample and its design, see Groves (1978). The University of Michigan Survey Research Center was not involved in implementation of the sample and has no responsibility for the final sample.

The sample contained 1,200 clusters of phone numbers for the U.S. The first two elements of telephone numbers are the area code and the central office code (AC-COC). The set of all valid combinations of AC-COC for the contiguous United States comprised the sampling frame. From these, a systematic sample of AC-COC combinations was obtained that reflected population density and geographical distribution strata. The sample contained 1,200 of the AC-COC combinations. For each combination, a four-digit number was randomly generated to complete the phone number. This number was designated the primary number representing a cluster of one hundred numbers in the same four-digit series. For example, if number 8305 was generated, the 8300–8399 series was used in the cluster. The other 99 numbers became designated as the secondary numbers in that cluster. The order in which the secondary numbers within a cluster would be called was also randomized.

Implementation

Each of the 1,200 primary numbers was called. If the number was found to be a nonworking number, that cluster was eliminated from the sample. There was a series of screening questions that eliminated all business phones and misdialed numbers from the sample. If the primary number was found to be a residential telephone number, the cluster was included in the sample.

Five attempts were made to reach someone at each primary number. Two attempts were made on different weekday evenings between 6 and 10 P.M. (All times given refer to local time at the number being called.) One attempt was made during the day (9 A.M. to 4 P.M.) on weekdays. Two attempts were made on weekends: one between noon and 5 P.M. and one after 5 P.M.

Valid clusters were identified in 318 cases. The number of households to be contacted in each valid cluster was set at 5. The same system of disposition codes, screening for business phones, and number and schedule of attempts was used for secondary numbers as was used for primary numbers. When a secondary number was established not to be a valid household number, it was replaced with another secondary number from that cluster.

Callback appointments were set up with respondents who were unable to complete the interview on first contact. As many as ten callbacks were made until a completion or refusal was established for that respondent.

Sex of respondent was designated on every primary and secondary number cover sheet to reflect approximately equal representation of

Appendix A

TABLE A.1
Approximate Sampling Tolerance (95 in 100 Confidence Interval)

	Percentage Near				
Size of Sample	10% or 90%	20% or 80%	30% or 70%	40% or 60%	50%
1,000	2%	3%	4%	4%	4%
500	3%	5%	5%	6%	6%
250	5%	6%	7%	8%	8%

males and females. If no one of the specified sex resided at the number, an interview was conducted with the resident adult.

Sampling Tolerance

The probability techniques used in this study attempt to represent the national population. However, survey results are subject to some variation because results are based on a sample rather than telephone interviews with the total population.

The size of the sample, variation among sample areas, and other factors influence sample variation. Table A.1 provides guidelines for interpreting confidence intervals for this study.

Comparisons with Census Data

Characteristics of the respondents approximate those of the general population. Most respondents were married (60 percent), owned homes (64 percent), were white (85 percent), lived in SMSA counties (76 percent), and were females (59 percent). Twenty-two percent of the respondents had total family incomes (before taxes) of less than $10,000, 26 percent had incomes of $10,000 to $19,000, 25 percent had incomes of $20,000 to $29,000, and 27 percent had incomes in excess of $30,000. These survey data correspond closely to the Bureau of the Census data as presented in Table A.2. For example, census data indicate that 58 percent are married, 83 percent are white, 75 percent of the population live in

SMSA counties, 60 percent are employed, 52 percent are females (19 years or older), and 28 percent have incomes under $10,000. The national survey differs somewhat from the census data on education, perhaps, reflecting the slightly higher proportion of women and those of upper-income levels in the sample.

TABLE A.2
Characteristics of Respondents from 1982 National Survey of
Adults (18 Years and Older) Compared with 1980 Bureau of Census Data

Characteristics	National Survey (N)	National Survey (percent)	Census (percent)
Sex			
Female	612	58.6	52.4
Male	432	41.4	47.6
R, NA, DK[a]	4	---	---
Race			
White	860	84.7 — white	83.3
Black	94	9.3	
Hispanic	36	3.5 nonwhite	
American Indian	12	1.2 15.3	16.7
Other	13	1.3	
R, NA, DK	33	---	
Tenure Status			
Own home	654	64.0	64.4
Rent	320	31.3	31.3
Other	48	4.7	4.3
R, NA, DK	26	---	
Marital Status			
Married	617	60.4	58.1[b]
Separated	25	2.4	2.4
Divorced	87	8.5	6.2
Widowed	74	7.2	7.3
Never married	219	21.4	25.8
R, NA, DK	26	---	
Employment Status			
Employed	641	62.7	60.4
Unemployed	77	7.5	4.3
Retired	121	11.8	
Homemaker	143	14.0 29.8	35.3
Student	41	4.0	
R, NA, DK	25	---	
Education			
Grade school	80	7.8	15.5
High school	524	51.3	53.2
College	323	31.6	25.3
Graduate degree	95	9.3	5.9
R, NA, DK	26	---	

TABLE A.2 (Continued)

Characteristics	National Survey (N)	National Survey (percent)	Census (percent)
Geographical Location			
Metropolitan	758	76.4	74.8
Nonmetropolitan	234	26.6	25.2
R, NA, DK	56	---	---
Place of Residence			
Farm	54	5.3 ⎱ 21.3	Rural 26.3
Rural nonfarm	163	16.0 ⎰	
Town	292	28.7 ⎱ 78.7	Urban 73.7
City	510	50.0 ⎰	
R, NA, DK	29	---	
Where Raised as a Youth			
Farm	190	18.6	
Rural	173	16.9	
Town	268	26.2	NA
City	391	38.3	
R, NA, DK	26	---	
Region			
Northeast	255	25.0	21.7
North Central	320	31.4	26.0
South	272	26.7	33.3
West	172	16.9	19.1
R, NA, DK	29	---	
Income (family)			
Under $5,000	84	8.4	11.8
$5,000 - $9,999	139	13.9	15.8
$10,000 - $19,999	263	26.3	28.2
$20,000 - $29,999	251	25.1	21.6
$30,000 - $39,999	128	12.8	11.8
$40,000 - $49,999	52	5.2	5.5
$50,000+	82	8.2	5.3
R, NA, DK	49	---	

[a] R=refused, NA=not applicable, DK=don't know.

[b] Includes 0.8% who indicated "other married, spouse absent."

TABLE A.2 (Continued)   Sources of Information

Sex — U.S. Bureau of the Census, Current Population Reports, Series P-25, No. 917, Table 2, Preliminary Estimates of the Population of the United States, by Age, Sex, and Race: 1970 to 1981. U.S. Government Printing Office, Washington, D.C.

Race — U.S. Bureau of the Census, PC80-S1-3, Table 2, Race of the Population by States: 1980. U.S. Government Printing Office, Washington, D.C.

Tenure Status — U.S. Bureau of the Census, Data User News, Vol. 17, No. 2, page 1, February, 1982. Department of Commerce, Washington, D.C.

Marital Status — U.S. Bureau of the Census, Population Characteristics, Series P-20, No. 372, page 7, Marital Status and Living Arrangements: March, 1981. U.S. Government Printing Office, Washington, D.C.

Employment Status — U.S. Bureau of the Census, Population Characteristics, Series P-20, No. 363, Table 21, Population Profile of the U.S.: 1980. U.S. Government Printing Office, Washington, D.C.

Education — U.S. Bureau of the Census, Population Characteristics, Series P-20, No. 356, page 40, Educational Attainment. U.S. Government Printing Office, Washington, D.C.

SMSA — U.S. Bureau of the Census, PC80-S1-5, Standard Metropolitan Statistical Areas and Standard Consolidated Statistical Areas: 1980, Table A. U.S. Department of Commerce, Washington, D.C.

Type of Residence — U.S. Bureau of the Census, Population Characteristics, Series P-20, No. 374, page 11, Table 2-4, Population Profile of the U.S.: 1981. U.S. Department of Commerce, Washington, D.C.

Region — U.S. Bureau of the Census, Statistical Abstract of the United States, 1981, page 12, No. 11. U.S. Department of Commerce, Washington, D.C.

Income — U.S. Bureau of the Census, Population Characteristics, Series P-20, No. 374, page 52, Table 9-3, Population Profile of the U.S.: 1981. U.S. Department of Commerce, Washington, D.C.

FIGURE A.1

NATIONAL SURVEY INSTRUMENT

PRIMARY NUMBER
COVER SHEET

Hello, my name is _____. I'm calling from the University of Kentucky in Lexington, Kentucky. Here at the University, we are working on a study for the Survey Research Center. First of all, I need to be sure that I dialed the right number.

Is this (telephone number)?

(IF NOT CLEAR) Since this telephone number has been generated by a computer, I do not know whether this number is for a business or a home. (Which is it?)

        1. Business   2. Home   3. Both

Does anyone live there on the premises?

    1. Yes   2. No

                END CONTACT

Do (they/you) have another phone number in the residence or do (they/you) use this number for personal calls?

    1. Have other   2. Use this

    END CONTACT

As I said, we are conducting this study from the University of Kentucky Survey Research Center. It is a national survey of people randomly selected throughout the United States on issues concerning food, agriculture, and quality of life. For most people, the survey will only take about 5 minutes. We have only a phone number and not any names so all answers are anonymous.

According to my instructions, I'm to talk to a (man/woman) at this number. Would that be you? (OR) Would you please call (him/her) to the phone? (BEGIN INTERVIEW OR REPEAT INTRODUCTION FROM "we are conducting..." IF NECESSARY.) (IF NO ONE OF SPECIFIED SEX LIVES IN THE HOUSEHOLD, INTERVIEW THE PERSON YOU ARE SPEAKING WITH. IF THE PERSON'S VOICE SOUNDS YOUNG, ASK IF THEY ARE OVER EIGHTEEN.)

                                                (INTERVIEW A
                                                Male   Female)

Call Record

Call Number	01	03	03	04	05
Date					
Time					
Disposition					
Interviewer					

```
 101   102 103 104 105 106 107
CARD     IDENTIFICATION NUMBER
```

If I have your permission, I would like to ask you some questions.

(Items 1 to 32 and 66 through 70 involved questions on food security, values, satisfaction with community life, and other issues not related to the Extension study.)

33. Have you ever heard of the Cooperative Extension Service (Sometimes called the agricultural Extension Service which is locally provided by County Extension Agents)?

```
              150 - 1 - No
                    2 - Yes
                    7 - N/A
                    8 - DK
                    9 - Refused
```

34. Have you ever heard of Extension Agricultural Programs?

    (Extension agricultural programs refer to any aspect of crop and livestock production and marketing, forestry, fisheries, wildlife and conservation. This includes such things as lawn and garden care, as well as farming.)

```
              151 - 1 - No
                    2 - Yes
                    7 - N/A
                    8 - DK
                    9 - Refused
```

35. Have you ever heard of Extension homemaker clubs or programs?

    (Extension homemaker clubs or programs refer to home economics programs in such areas as nutrition, clothing and textiles, family resource management, housing and home furnishings, and health.)

```
              152 - 1 - No
                    2 - Yes
                    7 - N/A
                    8 - DK
                    9 - Refused
```

36. Have you ever heard of Extension Community Development programs?

    (Extension Community Development programs focus on the solution of community problems such as the provision of services like fire protection, water and sewers; the expansion of businesses and industry; and the formation of local development organizations.)

```
              153 - 1 - No
                    2 - Yes
                    7 - N/A
                    8 - DK
                    9 - Refused
```

37. Have you ever heard of Extension 4-H youth programs?

(<u>Extension 4-H programs</u> stress the development of young people through projects, activities, and leadership development.)

154 - 1 - No --- | If <u>no</u> to all 5 questions, go to question 81.

2 - Yes - | If <u>yes</u> to 4-H question, continue.

If <u>yes</u> to any of 5 questions but <u>no</u> to 4-H question, <u>go</u> to question 40.

7 - N/A
8 - DK
9 - Refused

38. Were you a 4-H member as a youth?

155 - 1 - No
2 - Yes
7 - N/A
8 - DK
9 - Refused

39. Have you ever been a 4-H leader or helper?

156 - 1 - No
2 - Yes
7 - N/A
8 - DK
9 - Refused

40. Have you <u>personally ever</u> contacted an Extension agent or used the services of Extension?

157 - 1 - No
2 - Yes
7 - N/A
8 - DK
9 - Refused

41. Have other members of your family <u>ever</u> contacted an Extension agent or used services of Extension?

158 - 1 - No --- | If no to question 40 and 41 go to question 81.
2 - Yes
7 - N/A
8 - DK
9 - Refused

42. Within the past year have you personally contacted an Extension agent or used Extension services?

        159 - 1 - No
              2 - Yes
              7 - N/A
              8 - DK
              9 - Refused

43. Within the past year have other members of your family contacted an Extension agent or used services of Extension?

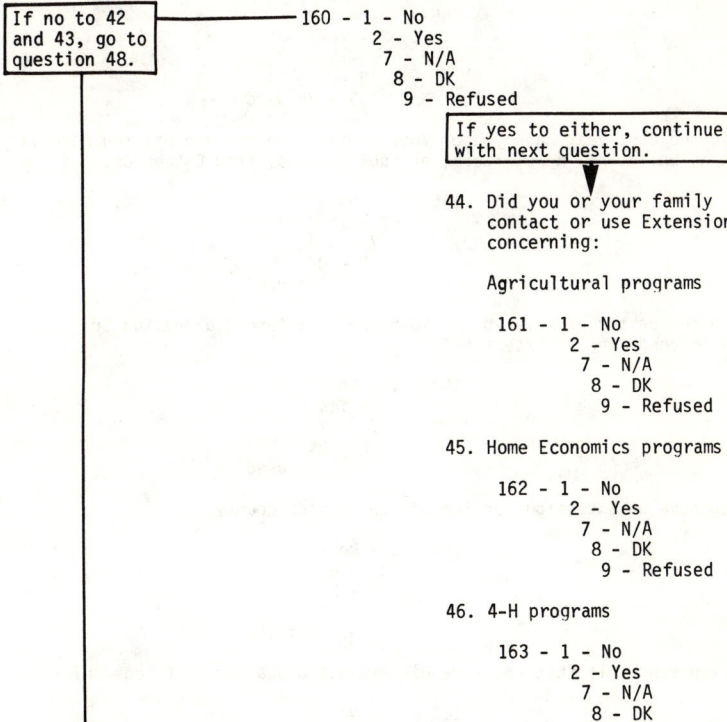

| If no to 42 and 43, go to question 48. |

160 - 1 - No
      2 - Yes
      7 - N/A
      8 - DK
      9 - Refused

If yes to either, continue with next question.

44. Did you or your family contact or use Extension concerning:

    Agricultural programs

    161 - 1 - No
          2 - Yes
          7 - N/A
          8 - DK
          9 - Refused

45. Home Economics programs

    162 - 1 - No
          2 - Yes
          7 - N/A
          8 - DK
          9 - Refused

46. 4-H programs

    163 - 1 - No
          2 - Yes
          7 - N/A
          8 - DK
          9 - Refused

47. Community Development
    programs

    164 - 1 - No
        2 - Yes
        7 - N/A
        8 - DK
        9 - Refused

48. Within the past year have you or your family listened to a radio program or watched a T.V. program conducted by Extension personnel?

    165 - 1 - No
        2 - Yes
        7 - N/A
        8 - DK
        9 - Refused

49. Within the past year have you or your family received any written material (such as bulletins, newsletters, or publications) from Extension?

    166 - 1 - No
        2 - Yes
        7 - N/A
        8 - DK
        9 - Refused

50. Within the past year have you of your family attended a meeting or workshop conducted by Extension?

    167 - 1 - No
        2 - Yes
        7 - N/A
        8 - DK
        9 - Refused

51. Do you have an Extension Service office in your county?

    168 - 1 - No
        2 - Yes
        7 - N/A
        8 - DK
        9 - Refused

52. Have you ever felt that you were discriminated against by Extension?

    169 - 1 - No
        2 - Yes
        7 - N/A
        8 - DK
        9 - Refused

    If yes, how? _____
    _____

The Extension Service is involved in many kinds of programs. Please indicate how high of a priority Extension should give to programs that focus on the following topics:(INTERVIEWER: REPEAT QUESTION AS NECESSARY)

Should a slight, moderate, great, or very great importance be given by Extension to:

		Slight	Moderate	Great	Very Great	NA	DK	RF
53. Food production	170 -	1	2	3	4	7	8	9
54. Energy conservation	171 -	1	2	3	4	7	8	9
55. Human nutrition	172 -	1	2	3	4	7	8	9
56. Community services and facilities	173 -	1	2	3	4	7	8	9
57. Family life and personal development	174 -	1	2	3	4	7	8	9
58. Housing	175 -	1	2	3	4	7	8	9
59. Economic Development	176 -	1	2	3	4	7	8	9
60. Consumer affairs	177 -	1	2	3	4	7	8	9
61. Home gardening and lawn care	178 -	1	2	3	4	7	8	9
62. Food marketing	179 -	1	2	3	4	7	8	9
63. Health care	180 -	1	2	3	4	7	8	9

```
                                        201   202 203 204 205 206 207
                                        CARD  IDENTIFICATION NUMBER
```

64. National resources and environment	208 -	1	2	3	4	7	8	9
65. Youth development	209 -	1	2	3	4	7	8	9

Are you very dissatisfied, dissatisfied, satisfied, or very satisfied with the following: (INTERVIEWER: REPEAT QUESTION AS NECESSARY)

		Very Dissatisfied	Dissatisfied	Satisfied	Very Satisfied	N/A	Don't Know	Refused
71. With the Cooperative Extension Service in general.	215 -	1	2	3	4	7	8	9
72. With Agriculture Extension programs	216 -	1	2	3	4	7	8	9
73. With 4-H Extension programs	217 -	1	2	3	4	7	8	9
74. With Home Economics Extension programs.	218 -	1	2	3	4	7	8	9
75. With Community Development Extension programs.	219 -	1	2	3	4	7	8	9

Now we would like to know whether you feel that less or more government funds should be spent on the following programs:

		Less	Same	More	N/A	DK	RF
76. Cooperative Extension Service in general.	220 -	1	2	3	7	8	9
77. Agriculture Extension programs.	221 -	1	2	3	7	8	9
78. Home Economics Extension programs	222 -	1	2	3	7	8	9
79. 4-H Extension programs.	223 -	1	2	3	7	8	9
80. Community Development Extension programs.	224 -	1	2	3	7	8	9

We would now like to ask you a few questions for background purposes. No individual responses can be identified.

81. Were you raised on a farm, in a rural area but not on a farm, in a town of less than 50,000, or in a city of 50,000 or more people?

    225 - 1 - Farm
    2 - Rural and Nonfarm
    3 - Town (less than 50,000)
    4 - City (50,000+)
    7 - N/A
    8 - DK
    9 - Refused

82. Do you now live on a farm, in a rural area but not on a farm, in a town of less than 50,000 people, or in a city of 50,000 or more people?

    226 - 1 - Farm
    2 - Rural and Nonfarm
    3 - Town (less than 50,000)
    4 - City (50,000+)
    7 - N/A
    8 - DK
    9 - Refused

83. In what state do you live? (write in) _____ 227 228

84. What is the name of your county? (write in) _____

    229 230 231 232

85. INTERVIEWER SKIP: SMSA
                        NonSMSA adjacent
                        NonSMSA nonadjacent   233

86. In what year were you born? (last two digits) _____ 234 235

87. What is the highest level of education that you have completed?

    236 - 1 - Grade school
    2 - High school
    3 - College
    4 - Graduate degree
    7 - N/A
    8 - DK
    9 - Refused

88. Are you currently employed?

    237 - 1 - Employed
    2 - Unemployed
    3 - Retired
    4 - Homemaker
    5 - Student
    7 - N/A
    8 - DK
    9 - Refused

89. Do you own or operate a farm?

238 - 1 - No
     2 - Yes --- If yes ---
     7 - N/A
     8 - DK
     9 - Refused

90. How many acres do you operate? _____

    239 240 241 242

91. Did your gross farm sales exceed $20,000 in 1981?

243 - 1 - No
     2 - Yes
     7 - N/A
     8 - DK
     9 - Refused

92. Is your operation primarily crop or livestock?

244 - 1 - Crop
     2 - Livestock
     3 - Both
     4 - Other
     7 - N/A
     8 - DK
     9 - Refused

93. In addition to farming, do you have an off-farm job?

245 - 1 - No
     2 - Yes
     7 - N/A
     8 - DK
     9 - Refused

94. If yes --- Is the job part-time or full-time?

246 - 1 - Part-time
     2 - Full-time
     7 - N/A
     8 - DK
     9 - Refused

95. Are you married, separated, divorced, widowed, or have you never been married?

    247 - 1 - Married
            2 - Separated
            3 - Divorced
            4 - Widowed
            5 - Never married
            7 - N/A
            8 - DK
            9 - Refused

96. Do you own your home, or do you rent?

    248 - 1 - Own home (or buying)
            2 - Rent
            3 - Live with parents/in-laws
            4 - Live with children/in-laws
            5 - Live with non-relatives
            7 - N/A
            8 - DK
            9 - Refused

97. Did you vote in the 1980 presidential election?

    249 - 1 - No
            2 - Yes
            7 - N/A
            8 - DK
            9 - Refused

98. Did you contribute money or your time to any political activities during the 1980 elections?

    250 - 1 - No
            2 - Yes
            7 - N/A
            8 - DK
            9 - Refused

99. Do you consider yourself politically . . .

    251 - 1 - Conservative
            2 - Middle-of-the-road
            3 - Liberal
            7 - N/A
            8 - DK
            9 - Refused

100. Which one of the following groups do you consider yourself a member of?
(INTERVIEWER: READ 1-6)

                        252 - 1 - White
                              2 - Black
                              3 - Hispanic (Puerto Rican,
                                            Mexican, etc.)
                              4 - American Indian
                              5 - Oriental
                              6 - Other _____
                              7 - N/A
                              8 - DK
                              9 - Refused

101. Finally, in 1981 was your total family income before taxes . . .
(INTERVIEWER: READ RESPONSE OPTIONS 1-7)

                        253 - 1 - Under $5,000
                              2 - $5 to $10,000 ($9,999)
                              3 - $10 to $20,000 ($19,999)
                              4 - $20 to $30,000 ($29,999)
                              5 - $30 to $40,000 ($39,999)
                              6 - $40 to $50,000 ($49,999)
                              7 - $50,000 or more
                              8 - N/A
                              9 - Refused

That's all of the questions that I have. Thank you very much for your time.

INTERVIEWER: COMPLETE THE FOLLOWING INFORMATION

GENDER OF THE RESPONDENT: 254 - 1 - Female
                                  2 - Male

TIME OF DAY: 255 - 1 - Morning
                                  2 - Afternoon
                                  3 - Evening

DATE OF COMPLETION:
                           $\overline{256}$ $\overline{257}$

TOTAL TIME OF INTERVIEW IN MINUTES:
                                     $\overline{258}$ $\overline{259}$

TELEPHONE AREA CODE:
                     $\overline{260}$ $\overline{261}$ $\overline{262}$

INTERVIEWER ID NUMBER:
                       $\overline{263}$ $\overline{264}$

Appendix B: Additional Findings

TABLE B.1
Awareness of Program Areas and Organizational Name by Personal Characteristics

Personal Characteristics	Agriculture (N=519) %	Home Economics (N=457) %	4-H (N=787) %	Community Development (N=462) %	Organizational Name (N=407) %
Age					
Less than 30	38	29	72	44	27
30-39	53	46	75	41	49
40-64	62	57	82	48	47
65 and over	56	50	81	53	38
	$x^2=39.0*$	$x^2=51.8*$	$x^2=11.8*$	$x^2=6.1$	$x^2=38.8*$
Sex					
Male	54	31	73	42	42
Female	50	54	80	48	39
	$x^2=.94$	$x^2=52.7*$	$x^2=6.9*$	$x^2=3.3$	$x^2=.65$
Race					
White	52	46	81	44	43
Black	52	43	57	47	26
Other	43	38	54	53	20
	$x^2=1.6$	$x^2=1.6$	$x^2=47.2*$	$x^2=2.0$	$x^2=21.4*$
Income					
Less than $10,000	43	42	68	48	33
$10,000-$19,999	56	49	82	48	39
$20,000-$29,999	49	38	76	41	36
$30,000 or more	55	50	82	43	53
	$x^2=10.3*$	$x^2=10.0*$	$x^2=16.1*$	$x^2=3.3$	$x^2=22.1*$
Education;					
Grade school	58	50	73	53	35
High school	47	42	76	45	38
College	54	45	80	46	42
Graduate School	63	57	83	1	55
	$x^2=11.0*$	$x^2=8.5*$	$x^2=4.5$	$x^2=2.4$	$x^2=10.7*$
Residence					
Farm	72	57	87	44	56
Rural nonfarm	61	57	86	47	54
Town	54	53	81	46	46
City	45	35	72	45	31
	$x^2=23.7*$	$x^2=33.5*$	$x^2=20.6*$	$x^2=.21$	$x^2=37.0*$
Farm Occupation					
No	49	43	77	45	39
Yes	79	65	85	53	61
	$x^2=22.6*$	$x^2=11.9*$	$x^2=2.5$	$x^2=1.7$	$x^2=13.2*$

* $p \leq .05$.

Table B.2
Profile of Knowledgeables of Extension

Personal Characteristics	Aware of Extension
	%
Income	
Less than $5,000	8
$5,000-$9,999	14
$10,000-$19,999	27
$20,000-$29,999	25
$30,000-$39,999	13
$40,000-$49,999	6
$50,000 or more	7
Education	
Grade school	8
High school	50
College	33
Graduate school	9
Age	
Less than 30	31
30-39	23
40-64	34
65 and over	12
Race	
White	86
Black	9
Other	5
Place of Residence	
Farm	6
Rural nonfarm	17
Town	30
City	47
Vote	
No	26
Yes	74
Political Contribution	
No	77
Yes	23
Political Orientation	
Conservative	36
Middle-of-the-road	42
Liberal	22
Farm Occupation	
No	92
Yes	8

TABLE B.3
Users and Nonusers of the Four Extension Program Areas in 1981

Select Characteristics of Users	Agriculture		Home Economics		4-H		Community Development	
	Nonuser (965)	User (88)	Nonuser (983)	User (61)	Nonuser (1004)	User (41)	Nonuser (1014)	User (30)
	%	%	%	%	%	%	%	%
Age								
30 or less	31	20	31	18	31	18	31	23
31-40	22	33	22	32	23	36	23	27
41-64	35	41	35	38	35	36	35	40
65 plus	12	6	11	12	11	10	11	10
	$x^2=11.7*$		$x^2=5.3$		$x^2=5.0$		$x^2=0.9$	
Sex								
Female	60	49	58	72	59	60	59	57
Male	40	51	42	28	41	40	41	43
	$x^2=3.3$		$x^2=4.3$		$x^2=0.0$		$x^2=0.0$	
Race								
White	84	93	84	86	85	85	85	79
Black	10	3	10	5	9	5	9	14
Other	6	3	6	9	6	10	6	7
	$x^2=5.3$		$x^2=1.83$		$x^2=1.96$		$x^2=0.8$	
Income								
Less than $10,000	24	7	23	14	23	8	23	10
$10,000-$19,999	27	27	26	34	26	37	26	38
$20,000-$29,999	25	26	25	27	26	13	25	28
$30,000 or more	24	39	25	25	25	42	25	24
	$x^2=16.0*$		$x^2=3.6$		$x^2=12.2*$		$x^2=3.5$	
Education								
Grade school	8	1	8	2	8	3	8	0
High school	52	41	52	48	52	28	52	40
College	32	31	31	35	31	38	32	33
Graduate school	8	26	9	15	8	31	9	27
	$x^2=37.3*$		$x^2=5.7$		$x^2=26.4*$		$x^2=13.2*$	
Residence								
Farm	4	18	5	10	5	18	6	0
Rural nonfarm	15	23	15	28	16	21	16	20
Town	29	21	28	32	28	33	29	30
City (50,000+)	51	38	51	30	51	28	50	50
	$x^2=38.6*$		$x^2=14.0*$		$x^2=16.9*$		$x^2=2.0$	
Occupation								
Farmer	5	30	7	18	7	28	7	13
Nonfarmer	95	70	93	82	93	72	93	87
	$x^2=66.2*$		$x^2=9.4*$		$x^2=22.4*$		$x^2=0.8$	
Gross Farm Sales								
Less than $20,000	91	65	83	75	87	54	82	80
$20,000 or more	9	35	17	25	13	46	18	20
	$x^2=5.7*$		$x^2=0.1$		$x^2=4.4*$		$x^2=0.0$	

* $p \leq .05$.

TABLE B.4
Satisfaction with Extension by Characteristics of the Respondent (N=251)

Personal Characteristics	Satisfied	x^2
	%	
Sex		
Male	96	
Female	95	1.50
Age		
Less than 30	92	
30-39	96	
40-64	95	
65 and over	94	5.99
Race		
White	95	
Nonwhite	94	.98
Income		
Less than $10,000	94	
$10,000 - $19,999	91	
$20,000 - $29,999	97	
$30,000 and over	98	.59
Education		
Grade school	90	
High school	95	
College	95	
Graduate degree	96	.99
Place of Residence		
Farm	96	
Rural nonfarm	92	
Town	96	
City	96	.43

* $p \leq .05$.

TABLE B.5
Satisfaction with Program Areas by Personal Characteristics

Personal Characteristics	Agriculture (N=238)	Home Economics (N=221)	4-H (N-230)	Community Development (N=206)
	%	%	%	%
Sex				
Male	95	96	96	83
Female	92	93	95	85
x^2	.29	.21	.65	.61
Age				
Less than 30	90	89	94	74
30-39	93	92	94	87
40-64	94	98	97	89
65 and over	94	88	95	73
x^2	6.36	14.12*	5.53	16.19*
Race				
White	92	93	95	84
Nonwhite	100	100	100	86
x^2	.28	.36	.65	.80
Income				
Less than $10,000	100	93	97	63
$10,000-$19,999	91	92	07	79
$20,000-$29,999	92	95	95	86
$30,000 and over	94	96	95	89
x^2	.44	.93	.42	.82
Education				
Grade school	91	89	82	78
High school	92	94	97	84
College	93	94	95	85
Graduate degree	95	95	95	83
x^2	.93	.96	.50	.66
Place of Residence				
Farm	96	100	96	95
Rural nonfarm	88	89	90	77
Town	93	92	97	82
City	95	97	98	88
x^2	.12	.06	.15	.54

* $p \leq .05$.

TABLE B.6
Support for Extension by Personal Characteristics

Personal Characteristics	Less	Same	More	x^2
	%	%	%	
Age				
Less than 30	20	33	47	
30-39	20	37	43	
40-64	15	51	34	
65 and over	21	47	32	5.75
Sex				
Male	19	45	36	
Female	16	39	45	2.08
Race				
White	18	44	38	
Nonwhite	12	25	63	3.78
Income				
Less then $10,000	17	45	38	
$10,000 - $19,999	15	42	43	
$20,000 - $29,999	21	37	42	
$30,000 and over	19	45	36	15.01*
Education				
Grade school	36	27	36	
High school	15	46	39	
College	21	37	42	
Graduate degree	15	48	37	5.36
Place of Residence				
Farm	13	61	26	
Rural nonfarm	17	41	42	
Town	17	53	30	
City	21	31	48	11.50*

* $p \leq .05$.

TABLE B.7
Support for Agricultural Programs by Personal Characteristics

	Less	Same	More	χ^2
	%	%	%	
Age				
Less than 30	6	17	77	
30-39	13	28	59	
40-64	7	48	45	
65 and over	24	35	41	24.08*
Sex				
Male	10	31	59	
Female	9	38	53	1.38
Race				
White	10	35	55	
Nonwhite	0	21	79	3.54
Income				
Less than $10,000	7	33	60	
$10,000 - $19,999	9	31	60	
$20,000 - $29,999	9	33	58	
$30,000 and over	12	38	50	13.35
Education				
Grade school	18	27	55	
High school	5	33	62	
College	16	27	57	
Graduate degree	7	55	38	17.38*
Place of Residence				
Farm	8	36	56	
Rural nonfarm	10	40	50	
Town	7	45	48	
City	12	23	65	10.62

* $p \leq .05$.

TABLE B.8
Support for Home Economics Progams by Personal Characteristics

Personal Characteristics	Less	Same	More	χ^2
	%	%	%	
Age				
Less than 30	22	47	31	
30-39	25	41	33	
40-64	16	51	33	
65 and over	31	31	38	4.27
Sex				
Male	21	48	30	
Female	20	44	36	0.98
Race				
White	23	46	31	
Nonwhite	0	40	60	7.13
Income				
Less than $10,000	14	43	43	
$10,000 - $19,999	19	38	43	
$20,000 - $29,999	23	45	32	
$30,000 and over	24	54	22	17.16*
Education				
Grade school	20	20	60	
High school	18	46	36	
College	23	42	35	
Graduate degree	24	60	16	11.44
Place of Residence				
Farm	13	65	22	
Rural nonfarm	21	42	37	
Town	21	49	30	
City	24	40	36	5.61

* $p \leq .05$.

TABLE B.9
Support for 4-H Programs by Personal Characteristics

Personal Characteristics	Less	Same	More	χ^2
	%	%	%	
Age				
Less than 30	15	38	47	
30-39	17	37	46	
40-64	8	49	43	
65 and over	16	31	53	5.78
Sex				
Male	10	41	49	
Female	14	41	45	1.09
Race				
White	13	43	44	
Nonwhite	7	20	73	4.87
Income				
Less than $10,000	10	37	53	
$10,000 - $19,999	11	40	49	
$20,000 - $29,999	14	40	46	
$30,000 and over	13	47	40	9.55
Education				
Grade school	18	27	55	
High school	10	43	47	
College	16	38	46	
Graduate degree	13	45	42	3.03
Place of Residence				
Farm	8	63	29	
Rural nonfarm	13	37	50	
Town	9	52	39	
City	16	31	53	13.22*

* $p \leq .05$.

TABLE B.10
Support for Community Development Programs by Personal Characteristics

Personal Characteristics	Less	Same	More	χ^2
	%	%	%	
Age				
Less than 30	13	37	50	
30-39	17	35	48	
40-64	15	48	37	
65 and over	21	26	53	5.95
Sex				
Male	18	43	39	
Female	15	37	48	1.53
Race				
White	17	41	42	
Nonwhite	0	25	75	7.43*
Income				
Less than $10,000	14	38	48	
$10,000 - $19,999	9	36	55	
$20,000 - $29,999	20	38	42	
$30,000 and over	18	45	37	17.47*
Education				
Grade school	36	18	46	
High school	13	42	45	
College	14	37	49	
Graduate degree	23	45	32	9.04
Place of Residence				
Farm	17	52	30	
Rural nonfarm	19	41	40	
Town	13	45	42	
City	16	31	53	6.49

*$p \leq .05$.

TABLE B.11
Correlation Coefficients for County-Level Variables in Kentucky (120)

County-Level Variables	X_1	X_2	X_3	X_4	X_5	X_6	X_7	X_8	X_9	X_{10}	X_{11}	X_{12}	X_{13}	X_{14}	X_{15}	X_{16}	X_{17}	X_{18}
X_1 - SMSA	1.00		Census Data	Co. Exp. Data						Data from the Extension Management Information System							Survey Data	
X_2 - Income from Agr (%)	.00	1.00																
X_3 - Income PC 1979[a]	-.60	-.19	1.00															
X_4 - Expenditures PC 1979	.26	.46	-.35	1.00														
X_5 - Staffdays PC[b]	.24	.40	-.32	.91	1.00													
X_6 - Contacts PC[b]	.16	.38	-.28	.70	.75	1.00												
X_7 - Office Contacts PC[b]	.03	.42	-.15	.73	.77	.54	1.00											
X_8 - Visits PC[b]	.33	.40	-.50	.80	.74	.57	.56	1.00										
X_9 - Telephone PC[b]	-.03	.44	.01	.49	.45	.43	.45	.32	1.00									
X_{10} - Workshops PC[b]	.02	.37	-.05	.40	.47	.53	.43	.33	.21	1.00								
X_{11} - Conferences PC[b]	.20	.09	-.14	.38	.37	.42	.12	.31	.13	.26	1.00							
X_{12} - Demonstrations PC[b]	.05	.13	-.16	.21	.12	.12	.07	.23	.06	-.01	.01	1.00						
X_{13} - Publications PC[b]	.03	-.01	.01	.18	.10	.14	.02	.12	.07	.03	.51	-.11	1.00					
X_{14} - Radio PC[b]	.22	.07	-.28	.27	.33	.25	.20	.35	.05	.12	.15	.17	.07	1.00				
X_{15} - Television PC[b]	.13	.14	.02	.22	.11	.02	.05	.14	.17	.04	.10	.04	.13	.07	1.00			
X_{16} - Use 1979 (%)	.25	.40	-.38	.61	.57	.53	.50	.51	.37	.37	.17	.07	.02	.15	.21	1.00		
X_{17} - Satisfaction 1979 (%)	.19	.45	-.21	.47	.46	.45	.44	.30	.38	.32	.21	.07	-.01	.16	.17	.64	1.00	
X_{18} - Support 1979 (%)	.11	.00	-.18	.05	.05	.13	.02	.03	.10	.04	.11	.08	.02	.15	-.10	.26	.14	1.00

[a]PC = per capita
[b]Three-year average, 1977-1979.

TABLE B.12
Number of County Staff Contacts (per capita) and Cost per Contact Regressed with Educational Methods and County Situation Variables

County Variables	Number of Contacts Per Capita, 1977-1979		Cost per Contact 1979	
	Without County Situation	With County Situation	Without County Situation	With County Situation
Office[a]	.16*	.18*	-.07	-.07
Visits[a]	.23*	.17*	.38*	.42*
Telephone[a]	.20*	.21*	-.14	-.13
Workshops[a]	.27*	.27*	-.19	-.17
Conferences[a]	.24*	.24*	-.20*	-.22*
Demonstrations[a]	.03	.03	-.10	-.09
Publications[a]	-.03	-.02	.14	.14
Radio[a]	.06	.05	-.02	-.02
Television[a]	-.08	-.06	.27*	-.13
SMSA		-.04		.06
Percent of Income from Agriculture		-.00		-.06
Income per Capita		-.12		.07
R^2	.56	.57	.23	.24

[a]Average yearly hours of county staff time, 1977-1979, expended on method (per capita).

* $p \leq .05$.

FIGURE B.1
Standardized Correlation Coefficients for County-Based Effectiveness Model

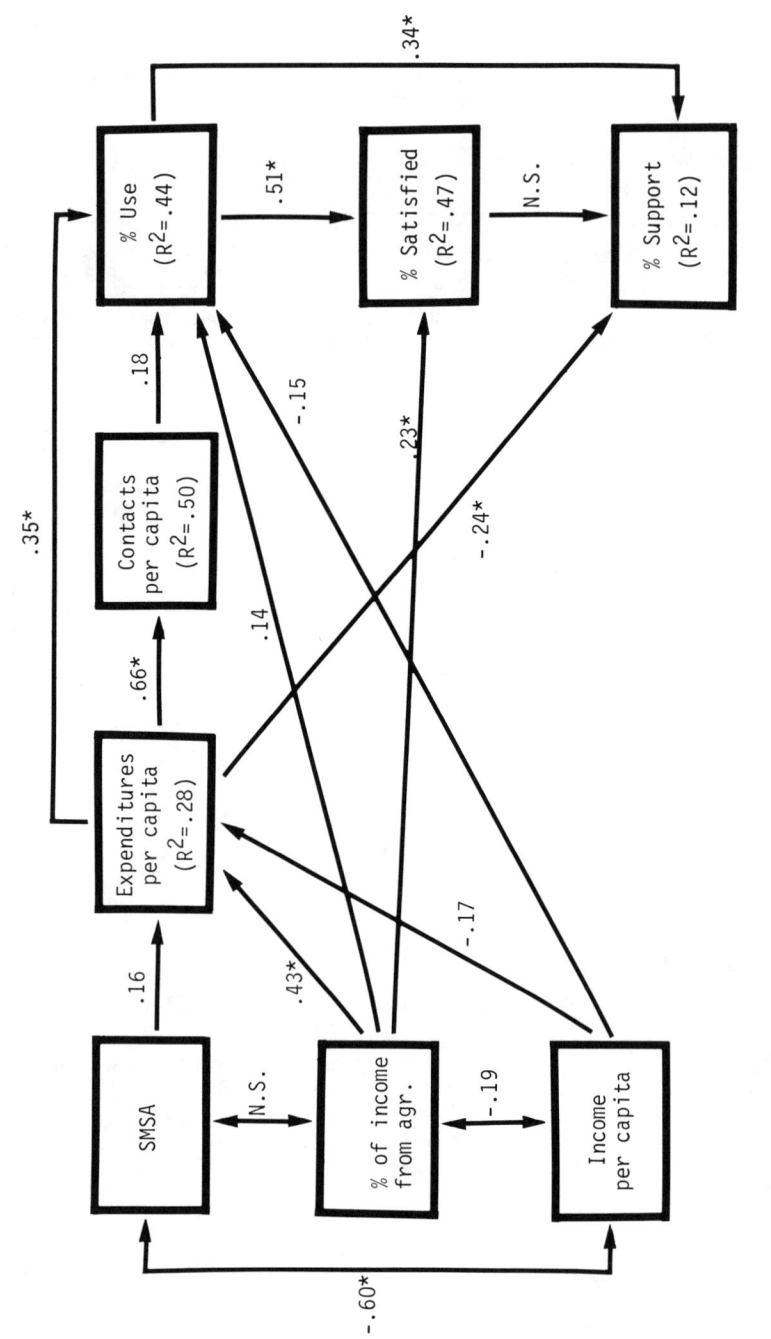

*p ≤ .05.   N.S. = Not Significant.

R^2 is calculated with all variables in the equation (relationships not shown had coefficients of ± .13 or less).

Index

Accountability. *See* Evaluation
Age. *See* Awareness; Clientele; Satisfaction/dissatisfaction; Support
Agriculture programs
  awareness, 48(figure), 49, 51, 54
  clientele, 36, 51, 59, 66, 67(figure), 68(table), 69, 119(table)
  contacts, 10(table)
  by education, 51, 68(table), 69
  by income, 51, 68(table), 69
  by occupation, 68(table)
  by place of residence, 51, 68(table), 69
  by race, 68(table)
  satisfaction, 77(figure)
  by sex, 68(table)
  staff, 12(table), 115, 119(table)
  support, 92(figure), 119(table)
  use, 66, 67(figure), 119(table)
Awareness, 33(figure), 35, 43, 45, 47, 49, 52–55, 115, 116(figure), 117, 136
  by age, 50, 51(table), 53
  of agriculture programs, 48(figure), 49, 51, 54
  by clientele, 48, 53
  of community development programs, 48(figure), 49, 52, 54
  by education, 53
  extent, 47–49
  of 4-H programs, 48(figure), 49, 52, 54, 115, 119(table), 136
  of home economics programs, 48(figure), 49, 51, 52, 54
  by income, 50, 51(table), 53
  name recognition, 52
  by place of residence, 50(table), 53
  by race, 50, 52(table), 53
  by region, 49(figure)
  See also Cooperative Extension Service, image; Systems Effectiveness Model

Baker, Ron, 132
Bennett, Claude F., 18, 19
Budget. *See* Expenditures; Funding; Support

CES. *See* Cooperative Extension Service
Clientele, 21–22, 35–36, 44, 56–58, 66, 87, 89, 134, 137–138
  by age, 63, 64(table)
  in agriculture programs, 36, 51, 59, 66, 67(figure), 68(table), 69, 119(table)
  in community development programs, 52, 59, 66, 67(figure), 68(table), 69, 71, 117, 119(table), 138
  compared with nonusers, 65(table), 67(table), 78(table), 79, 81
  contacts, 58, 69, 70, 71(table), 72, 112–113

189

discrimination, 63–64, 66, 71, 117, 118(table), 138–139
by education, 65(table), 66, 68(table)
in 4-H programs, 35, 52, 59, 66, 67(figure), 68(table), 69, 119
frequency of use, 69, 70(table), 81, 83(table)
in home economics programs, 51, 52, 59, 66, 67(figure), 68(table), 69, 120(table)
household use, 59, 60(table), 61(figure), 70–72, 117, 118(table), 137
by income, 65(table), 66, 68(table)
individual use, 59, 60(table), 61
by marital status, 65(table), 66
by occupation, 62(table), 63, 68(table)
by place of residence, 21, 61, 62(table), 63, 65(table), 68(table), 69, 70–71, 117, 119(table), 137–138
by political orientation, 66, 67(table)
by race, 64(table), 66, 68(table), 118(table), 138
by region, 61(figure), 71
by rural background, 62(table), 63, 69, 70(table), 71, 119(table)
satisfaction and use, 37, 75(table), 76, 78(table), 79, 80(table), 81, 83(table), 117, 119(table), 120, 139–140
by sex, 63, 64(table), 68(table), 118(table)
in Smith-Lever Act, 6, 7, 8, 32, 34, 57
support, 22–23, 37–39, 91, 92(figure), 93(figure), 94(figure, table), 95(table), 96(table), 97(table), 98
Communication methods, 23–24, 54, 69, 70, 71(table), 109, 117, 118(table), 140–142
group, 23, 70, 71(table), 104, 105, 107, 108(table), 109, 140, 141
individual, 23, 54, 104, 105, 107, 108(table), 109, 118(table), 141
in Kentucky, 104, 105, 107, 108(table), 109, 141
mass media, 54, 70, 71(table), 104, 105, 107, 108(table), 109, 118(table), 138
See also Contacts
Community development programs
awareness, 48(figure), 49, 52, 54
clientele, 52, 59, 66, 67(figure), 68(table), 69, 71, 117, 119(table), 138
contacts, 10(table)
by education, 52, 68(table)
by income, 52, 68(table)
by occupation, 68(table)
by place of residence, 52, 68(table), 69
by race, 52, 68(table), 138
satisfaction, 77(figure)
by sex, 68(table)
staff, 12(table), 115, 119(table)
support, 92(figure)
use, 66, 67(figure)
Constituency. *See* Clientele
Contacts, 58, 117, 118(table)
in agriculture programs, 10(table)
in community development programs, 10(table)
cost, 109–110, 112–113
in 4-H programs, 10(table), 119(table)
group, 23, 112
in home economics programs, 10(table)
individual/personal, 23, 112–113
Kentucky data, 107, 109, 112–113
Cooperation Extension Service (CES), 118(table)
goals, 34, 54
history, 5–9, 58, 133, 137. *See also* Smith-Lever Act

identity, 20–21, 47, 52, 53–55, 88, 135–136
image, 43, 44–47, 53–55, 117, 118(table), 119(table), 135–136
mission, 18, 24, 48, 57, 118, 125–128
role, 5, 9, 57, 118(table), 135, 142
stereotypes, 20–24
structure, 5, 11, 129–130
Cost-bearer, 44, 78, 84, 145
County agents. *See* Staff
County programs. *See* Agriculture programs; Community development programs; 4-H programs; Home economics
Cronback, Lee J., 26
Cummings, Larry L., 29, 73

DeMarco, Susan, 10, 12, 88, 102
Discrimination. *See* Clientele
Dissatisfaction. *See* Satisfaction/dissatisfaction

Earmarked funds. *See* Funding
ECOP. *See* Extension Committee on Organization and Policy
Education. *See* Agriculture programs; Awareness; Clientele; Community development programs; 4-H programs; Home economics programs; Nonusers; Satisfaction/dissatisfaction; Support
Education methods. *See* Communication methods
Effectiveness. *See* Evaluation
EMIS. *See* Extension Management Information System
Environment, 39–40, 133–135
Equal opportunity. *See* Clientele, discrimination
Etzioni, Amitai, 28
Evaluation, 41–42, 124, 144–146
client-based, 145
effectiveness, 26–29, 36, 41, 73, 86, 144, 146

efficiency, 26
goal approach, 27, 28–29, 74
Miller's Three-Stage Model, 29–30(figure)
objective measures, 73–74
subjective measures, 73–74
systems approach, 27, 29
See also Food and Agriculture Act (1977); Extension Committee on Organization and Policy; General Accounting Office; Systems Effectiveness Model
Expenditures
Kentucky data, 110, 111(figure), 112–113
See also Funding; Inputs
Extension agent. *See* Staff
Extension Committee on Organization and Policy (ECOP), 2, 17, 18, 24, 132
Extension Management Information System (EMIS), 2, 3, 4, 16, 30, 34, 58, 73, 74, 100, 102, 104, 113, 117, 145
compared with survey estimates, 107, 109

Farm size. *See* Satisfaction/dissatisfaction
Food and Agriculture Act (1977), 2, 17
Formula funds. *See* Funding
4-H programs
awareness, 48(figure), 49, 52, 54, 115, 119(table), 136
clientele, 35, 52, 59, 66, 67(figure), 68(table), 69, 119(table)
contacts, 10(table), 119(table)
by education, 68(table), 69
by income, 52, 68(table), 69
by occupation, 68(table)
by place of residence, 52, 68(table)
by race, 52, 54, 68(table)
satisfaction, 77(figure)
by sex, 68(table)

staff, 12(table), 14, 115
support, 92(figure)
use, 35, 63, 66, 67(figure), 119(table)
volunteers, 14
Funding, 14–16, 15(figure), 16, 39
county, 15(figure), 16
earmarked, 9, 16, 39, 131
federal, 15(figure), 16, 130–133
formula, 16, 39, 131–132
state, 15(figure),16, 130
See also Inputs; Support; Systems Effectiveness Model; Systems Model

GAO. *See* General Accounting Office
General Accounting Office (GAO), 2, 24, 78
mission statement, 18
Goals. *See* Cooperative Extension Service; Evaluation, goal approach

Hildreth, R. J., 9
Home economics programs
awareness, 48(figure), 49, 51, 52, 54
clientele, 51, 52, 59, 66, 67(figure), 68(table), 69, 119(table)
contacts, 10(table)
by education, 52, 68(table)
by income, 52, 68(table)
by occupation, 68(table)
by place of residence, 51, 68(table)
by race, 68(table)
satisfaction, 77(figure)
by sex, 68(table), 119(table)
staff, 12(table), 14, 115, 120(table)
support, 92(figure), 119(table)
use, 66, 67(figure)
volunteers, 14

Identity. *See* Cooperative Extension Service

Image. *See* Awareness; Cooperative Extension Service
Income. *See* Awareness; Agriculture programs; Clientele; Community development programs; 4-H programs; Home economics programs; Nonusers; Satisfaction/dissatisfaction; Support
Inequity. *See* Clientele, discrimination
Inputs, 33(figure), 34, 116(figure). *See also* Systems Effectiveness Model; Systems Model

Jenkins, John W., 22

Katz, Daniel, 2, 37, 43, 56, 75(table), 76, 90
Kentucky data, 40, 41
communication methods, 107, 109
cost effectiveness, 109–110
county size, 106–107
expenditures, 107
frequency of use, 69, 70(table), 81, 83(table)
income, 107
Kentucky Extension Management Information System, 40
satisfaction, 78(table), 79, 80(table), 81, 82(table), 83(table), 105, 111(figure), 112
support, 96, 97(tables), 98, 106, 111(figure), 112
survey method, 40, 78, 101, 106
See also Systems Effectiveness Model
Knapp, Seaman, 7, 23

Legitimacy/legitimization, 19–20, 24, 29, 39, 41, 136
Loomis, Charles P., 100

Marital status. *See* Clientele; Nonusers

Index

Mass media. *See* Communication methods
Mendelsohn, Harold, 46
Metropolitan/nonmetropolitan residence. *See* Clientele, by place of residence
Meyer, Marshall W., 19

Naisbitt, John, 130
Name recognition. *See* Awareness; Cooperative Extension Service, identity, image
National Agricultural Research and Extension Users Advisory Board, 56
Nonusers, 119(table), 140
  by age, 81, 82(table)
  compared with clientele, 65(table), 67(table), 78(table), 79, 81
  dissatisfaction, 81, 83, 119(table), 120
  by education, 65(table), 81, 82(table)
  by farm sales, 82(table)
  by farm size, 82(table)
  by income, 65(table), 81, 82(table)
  in Kentucky, 81, 82(table)
  by marital status, 65(table)
  by place of residence, 81, 82(table)
  by political orientation, 67(table)
  by race, 81
  by rural background, 119(table)
  satisfaction, 78(table), 81, 82(table), 83(table), 119(table), 120
  by sex, 82(table)
  support, 37, 96(table), 97(tables), 98, 142
  See also Systems Effectiveness Model

Occupation. *See* Agriculture programs; Awareness; Clientele; Community development programs; 4-H programs; Home economics programs; Satisfaction/dissatisfaction
Outputs. *See* Systems Effectiveness Model; Systems Model
Oversight hearings. *See* U.S. House of Representatives

Place of residence. *See* Agriculture programs; Awareness; Community development programs; 4-H programs; Home economics programs; Nonusers; Satisfaction/dissatisfaction; Support
Political orientation. *See* Clientele; Nonusers
Population
  farm, 7, 8(figure), 47, 132, 134
  part-time farm, 134
Priorities
  for future, 120–122, 123(table), 124
Program operations. *See* Systems Effectiveness Model; Systems Model
Public opinion, 44–46, 88–89

Race. *See* Agriculture programs; Awareness; Clientele, discrimination; Community development programs; 4-H programs; Home economics programs; Nonusers; Satisfaction/dissatisfaction; Support
Region. *See* Awareness; Clientele; Satisfaction/dissatisfaction; Support
Rural/urban. *See* Clientele

Satisfaction/dissatisfaction, 22, 36–37, 43, 84–85, 116(figure)
  by age, 77, 79, 80(table), 82(table)
  agriculture programs, 77(figure)

community development
  programs, 77(figure)
by education, 77, 79, 80(table),
  82(table)
by farm sales, 80(table), 82(table)
by farm size, 79, 80(table),
  82(table)
4-H programs, 77(figure)
by frequency of use, 81, 83(table)
with government agencies,
  75(table)
home economic programs,
  77(figure)
by income, 77, 79, 80(table),
  82(table)
Kentucky data, 78(table), 79,
  80(table), 81, 82(table),
  83(table), 105, 111(figure), 112
of nonusers, 78(table), 81,
  82(table), 83(table), 119(table),
  120
by occupation, 77
by place of residence, 77, 79,
  80(table), 81, 82(table)
by race, 77, 80(table), 81
by region, 76(figure)
by sex, 77, 80(table), 81, 82(table)
and support, 37–38, 90, 95(table),
  96(table), 99, 111(figure), 142
and use, 75(table), 76, 78(table),
  79, 80(table), 81, 82(table),
  83(table), 111(figure), 112, 117,
  119(table), 120, 139–140, 142
See also Clientele; Systems
  Effectiveness Model
Scott, W. Richard, 74
Sex. See Agriculture programs;
  Clientele; Community
  development programs; 4-H
  programs; Home economics
  programs; Nonusers;
  Satisfaction/dissatisfaction
Smith, Clarence B., 22
Smith-Lever Act, 6, 7, 8, 32, 39, 57
Specialists. See Staff, state
Staff, 101, 118(table), 121(table), 122

agriculture programs, 12(table),
  115, 119(table)
community development
  programs, 12(table), 115,
  119(table)
contacts, 23–24, 112–113
county, 9, 11(table), 12, 13(table),
  20, 23, 104, 134
federal, 10, 11(table)
4-H programs, 12(table), 14, 115
growth rate, 115
home economics programs,
  12(table), 14, 115, 119(table)
methods, 107, 108(table)
size, 104, 129
specialization, 12, 14, 129–130
staffdays, 104
state, 11(table), 12, 13(table)
State studies
  Kentucky. See Kentucky data
  Missouri, 60
  Oklahoma, 60
  Wisconsin, 60
Stereotypes. See Cooperative
  Extension Service
Stockdale, Jerry D., 134
Suchman, Edward A., 31
Support, 22–23, 37–38, 43, 88–90,
  116(figure), 119(table), 140,
  142–144
agriculture programs, 92(figure),
  119(table)
community development
  programs, 92(figure)
by education, 95, 98
4-H programs, 92(figure)
by frequency of use, 97(table)
by general public, 37–38, 98–99,
  143
history, 90–91, 99
home economics programs,
  92(figure), 119(table)
by income, 95, 98
Kentucky data, 96, 97(table), 98,
  106, 111(figure), 112

national survey, 91, 92(figure),
  93(figure), 94(figure, table),
  95(table)
nonusers, 37, 96(table), 97(tables),
  98, 142
organized, 38–39
by place of residence, 95, 98, 99
by race, 95, 98
by region, 93(figure), 94(table)
and use, 91, 92(figure), 93(figure),
  94(figure, table), 95(table),
  96(table), 97(tables), 98, 99,
  111(figure), 112, 119(table), 120,
  142
See also Cooperative Extension
  Service; Clientele; Systems
  Effectiveness Model
Systems Effectiveness Model, 4, 32,
  33(figure), 42, 100, 103(figure),
  111(figure), 115, 116(figure), 144
  awareness, 33(figure), 35,
    116(figure)
  county, 100, 102, 103(figure),
    104–110, 111(figure), 112–114
  data sources, 40
  environment, 39–40, 102,
    103(figure), 106, 111(figure),
    116(figure)
  funding, 39
  impact, public, 102
  inputs, 32, 33(figure), 34, 39, 102,
    103(figure), 104, 116(figure),
    118(table)
  nonusers, 35
  organization, 32, 33(figure), 34,
    116(figure)
  outputs, 32, 33(figure), 36,
    103(figure), 105–106
  program operations, 32,
    33(figure), 34, 102, 103(figure),
    104–105, 116(figure)
  satisfaction, 33(figure), 36–37,
    111(figure), 116(figure)
  support, 33(figure), 37–39,
    111(figure), 116(figure)

users, 35–36, 105, 111(figure),
  116(figure)
Systems Model, 28(figure), 29

Telecommunications. *See*
  Communication methods;
  Contacts
Tripodi, Tony, 30
True, Alfred C., 38

USDA-NASULGC, 5, 132
Use, 35–36, 43, 58–59, 105, 112,
  116(figure), 118(table), 137–138,
  139–140
  household, 59, 60(table),
    61(figure), 70–72, 117,
    118(table), 137
  individual, 59, 60(table), 61
  priorities by program, 122,
    123(table), 124
  and satisfaction/dissatisfaction,
    37, 75(table), 76, 78(table), 79,
    80(table), 81, 82(table),
    83(table), 95, 96(table),
    111(figure), 112, 117, 119(table),
    120, 139–140, 142
  and support, 91, 92(figure),
    93(figure), 94(figure, table),
    95(table), 96(table), 97(tables),
    98, 99, 111(figure), 112,
    119(table), 120, 142
  See also Agriculture programs;
    Clientele; Community
    development programs; 4-H
    programs; Home economics
    programs; Systems
    Effectiveness Model
Users Advisory Board. *See* National
  Agricultural Research and
  Extension Users Advisory
  Board
U.S. House of Representatives
  oversight hearings, 18, 57, 132

Wholey, Joseph S., 41

Yuchtman, Ephraim, 86